I0475757

TABLES

FIGURE

BOXES

SUMMARY
_____

_____

_____

_____

_____

Unmanned aerial vehicles (UAVs) are remotely piloted or self-piloted aircraft that can carry cameras, sensors, communications equipment, or other payloads. The Department of Defense has used UAVs in military operations since the 1950s because they can provide reconnaissance, surveillance, and intelligence of enemy forces without risking the lives of an aircrew. In recent years, interest in the many capabilities of UAVs has been growing among the armed services. At the same time, the services have been having difficulty actually acquiring and deploying the UAVs they have tried to develop. As a result, many of those development programs have been cancelled (see Summary Table 1). The Department of Defense (DoD) hopes to do better with the four UAVs that are now under development or in initial production: Predator, Darkstar, Global Hawk, and Outrider.

In this paper, the Congressional Budget Office (CBO) examines DoD's unmanned aerial vehicle programs, reviewing their missions, requirements, and development process. It also focuses on the strengths and weaknesses of the Advanced Concept Technology Demonstration (ACTD) process being used to develop them. Criticisms of the new unmanned aerial vehicles have led some Members of Congress to question whether reasonable alternatives exist to DoD's plans for acquiring and using UAVs. This paper examines five options for the various UAV programs. Those alternatives illustrate other potential configurations of the future UAV force—each of which would have advantages and disadvantages that would differ from those of the current combination of programs.

PROMISE AND PROBLEMS IN DoD'S UAV PROGRAMS

Unmanned aerial vehicles make up a small part of the defense budget (DoD currently spends about $600 million a year on all UAV acquisition programs—less than 1 percent of its acquisition budget). Nevertheless, they hold great promise. Military thinkers who contend that warfare is becoming more information-based believe that UAVs can play a key role by providing their users with sustained, nearly instantaneous video and radar images of an area without putting human lives at risk. At the tactical level—the local area of operations—that is a capability that battalion and brigade commanders have not had before. At the longer-range, strategic level, UAVs have some advantages over reconnaissance satellites, such as being able to watch one area for an extended period of time. (Eventually, unmanned aerial vehicles may also be used in combat operations, such as the suppression of enemy air defenses and strike missions, but those developments are still years from realization.)

Because the different military services have various imagery needs at various levels, DoD plans to field a family of UAV systems. The only one currently

SUMMARY TABLE 1. MAJOR UNMANNED AERIAL VEHICLE PROGRAMS

Program	Period	Description	Status
Lightning Bug	1964-1979	Reconnaissance drone first used by the Air Force during the Vietnam War	Retired
Aquila	1979-1987	Tactical UAV for Army commanders	Canceled
Amber	1984-1990	Classified endurance UAV	Canceled
Pioneer	1986-present	UAV originally acquired to assess battle damage by naval gunfire	Deployed
Medium Range	1987-1993	Tactical UAV for the Air Force and Navy	Canceled
Hunter	1988-1996	Joint tactical UAV	Canceled after LRIP[a]
Gnat-750	1988-present	Long-endurance UAV developed with CIA funding; exported commercially	Used for training and intelligence missions
Predator	1994-present	Long-endurance UAV for theater commanders; based on the Gnat-750	In LRIP
Darkstar	1994-present	Stealthy endurance UAV for high-threat environments	In development
Global Hawk	1994-present	High-altitude, long-range endurance UAV	In development
Outrider	1996-present	Joint tactical UAV	In development

SOURCE: Congressional Budget Office.

NOTE: UAV = unmanned aerial vehicle; LRIP = low-rate initial production; CIA = Central Intelligence Agency.

a. Seven systems—each equipped with eight air vehicles, four ground control stations, and support equipment—were initially placed in storage. Later, most of the equipment for two systems was re moved and used in training exercises and in developing "concepts of operation" for UAVs.

deployed with U.S. troops is Pioneer, which the Navy and Marine Corps use for a variety of tactical operations, including surveying potential targets and assessing damage from naval missiles and gunfire. Pioneer, which is now 12 years old, is considered outmoded and is due to be retired in 2003. That date may be pushed back a few years, however, while a successor system is developed.

The system that is next-farthest along is Predator, a medium-altitude UAV that has finished its development phase and entered low-rate initial production. Predator is an Air Force system designed to meet the reconnaissance needs of the theater commander in a major regional conflict. The Army, however, would also like its corps and division commanders to be able to use Predator in such conflicts.

The three newest UAVs—Darkstar, Global Hawk, and Outrider—are still at the development stage. Darkstar and Global Hawk, which are being developed together, are both high-altitude UAVs that would be used (in different ways) by theater commanders and national command staff for reconnaissance and surveillance. Outrider, in contrast, is a tactical UAV geared toward brigade and task-force commanders.

In reviewing those unmanned aerial vehicle programs, CBO has identified three issues of concern. Two of those issues relate to the tactical UAVs that the services would put into the field. The third relates to the high-altitude UAVs.

First, Outrider (the tactical UAV under development) may not be suitable for all of its intended missions. Although it is being developed as a joint program for the Army, Navy, and Marine Corps, Outrider will not have several attributes that are important for a Navy or Marine system operating from ships: vertical take-off and landing capability and an engine that can run on heavy fuel such as jet or diesel fuel. (It should be noted that the technical objectives of the Outrider program included a heavy-fuel engine, which was subsequently scrapped, but did not include vertical take-off and landing capability.)

Second, the manner in which the Army is planning to fulfill its corps- and division-level UAV requirements during a regional conflict may not prove feasible. The Army intends to rely on the Predators being bought and operated by the Air Force (at least as of this writing). However, Predator will receive its assignments from the theater commander; thus, the needs of Army field commanders for information that UAVs are intended to provide are not likely to be met if other missions receive higher priority.

Third, there may be overlaps in the capability provided by Predator, Global Hawk, and Darkstar. By various performance measures, such as speed, operating altitude, payload, and range, Darkstar falls between the other two (see Summary

Table 2). But Darkstar is intended to be more survivable than either of them because it will have stealth characteristics. However, it is not clear whether a substantial number of stealthy unmanned aerial vehicles are necessary. If not, Predator and Global Hawk might be able to perform many of Darkstar's intended missions.

DEVELOPING UNMANNED AERIAL VEHICLES WITH THE ACTD PROCESS

Developing unmanned aerial vehicles has not been easy. Technical challenges and growth of costs have led to a number of unsuccessful UAV programs. Unlike those past efforts, however, the UAVs under development today are using a different process, called Advanced Concept Technology Demonstration. ACTDs are supposed to be small-budget, low-risk demonstrations of a new technology and are usually focused on meeting a specific requirement identified by the technology's intended users. The purpose of having an ACTD program that is separate from DoD's traditional acquisition and development process is to give developers a flexible

SUMMARY TABLE 2.   COMPARISON OF THE CAPABILITIES OF PREDATOR, DARKSTAR, AND GLOBAL HAWK

	Predator	Darkstar	Global Hawk
Maximum Range (Kilometers)[a]	4,200	5,600	25,500
Operating Altitude (Feet)	10,000-25,000	40,000-45,000	55,000-65,000
Cruise Speed (Kilometers per hour)	120	463	639
Endurance at Radius	20 hours at 926 km	8 hours at 926 km	22 hours at 5,556 km
Payload (Pounds)	450	1,000	2,000
Survivability	?	?	?

SOURCE: Congressional Budget Office based on data from the Department of Defense.

NOTE: km = kilometers.

a.   Maximum range is the farthest the unmanned aerial vehicle (UAV) can fly before running out of fuel. It differs from radius in that the tactical UAVs are limited to an operating radius far short of the range of the air vehicle because they communicate through line-of-sight links. Endurance UAVs are not limited to a particular radius because, when not in an autonomous mode, they communicate with their controllers by satellite.

management environment in which to experiment freely with new technologies and demonstrate their utility on the battlefield to commanders.

The current UAV development programs, however, are technologically ambitious. The Outrider program and the combined Global Hawk/Darkstar program have experienced numerous problems, including delayed schedules, growing costs, and the crash of a Darkstar. Predator, by contrast, moved relatively smoothly from the ACTD stage to low-rate initial production, perhaps because its contractor had an operational predecessor to work with as well as experience in integrating the many components that make up a working UAV system. Outrider, Global Hawk, and Darkstar represent much more difficult development projects. The troubles they have experienced are not atypical of acquisition programs, but those troubles come as a disappointment to people who expected ACTDs to be affordable and relatively quick demonstrations of proven technology.

ILLUSTRATIVE OPTIONS FOR DoD'S UAV PROGRAMS

The Congressional Budget Office has constructed five options to address the concerns that have been raised about DoD's UAV development efforts or to take greater advantage of the promise that UAVs appear to hold. Each option deals with a particular problem or aspect of the way the services or the Department of Defense are planning to develop, acquire, and use UAVs. Because they address only one particular issue of the UAV programs and missions, the options are not comparable with one another. Nor do they represent all of the possible ways to improve DoD's UAV programs; there are others that CBO did not consider. Moreover, the options were designed to address specific problems, not to generate savings.

Because many of the UAVs discussed in this paper are still in the ACTD phase and DoD has not yet committed to buying them in quantity, these options cannot be compared with an overall Administration plan. Decisions about acquisition must wait until the end of the UAVs' development and demonstration process. In the absence of concrete plans by the services or DoD to purchase particular UAVs in specific quantities, CBO compared its options—in terms of both cost and capability—with its assumption of what the services or DoD will eventually buy, based on information they provided. Furthermore, that comparison is predicated on the assumption that the UAVs now under development—Darkstar, Global Hawk, and Outrider—will all be ultimately successful and DoD will buy and deploy them.

CBO's options vary widely in their potential costs or savings (see Summary Table 3). The total for each option represents acquisition costs as well as operating and support costs over the assumed 15-year life of the UAVs. Estimates of operating

SUMMARY TABLE 3.   COSTS AND SAVINGS FOR FIVE ILLUSTRATIVE OPTIONS
FOR UAVs (In millions of 1998 dollars)

	Acquisition Costs	15-Year Operating and Support Costs	Total
Option I: Provide a UAV Capability to Brigade and Task-Force Commanders			
DoD's Plan[a]	860	930	1,790
Cost of Option IA	780	1,020	1,800
Cost or Savings (-) Compared with DoD's Plan[b]	-80	90	10
Cost of Option IB	640	1,010	1,650
Cost or Savings (-) Compared with DoD's Plan	-220	80	-140
Option II: Provide a UAV Capability to Army Corps and Divisions Commanders			
Army's Plan[c]	0	0	0
Cost of Option II	250	500	750
Cost Compared with Army's Plan	250	500	750
Option III: Trade Off UAVs for Reconnaissance Helicopters			
Army's Plan[d]	31,500	6,600	38,200
Cost of Option III	27,700	6,000	33,700
Savings Compared with Army's Plan	-3,800	-700	-4,500
Option IV: Supplement JSTARS Coverage with UAVs			
Air Force's Plan	1,700	4,300	6,000
Cost of Option IV	2,200	5,000	7,200
Cost Compared with Air Force's Plan	500	700	1,200
Option V: End Darkstar Production with the ACTD Vehicles			
Air Force's Plan[a]	2,600	1,900	4,600
Cost of Option V	2,000	1,600	3,600
Savings Compared with Air Force's Plan	-600	-400	-1,000

SOURCE: Congressional Budget Office.

NOTE: UAV = unmanned aerial vehicle; DoD = Department of Defense; JSTARS = Joint Surveillance Target Attack
Radar System; ACTD = Advanced Concept Technology Demonstration.

a.   CBO's assumed plan based on available information.

b.   The new UAV for the Navy and Marine Corps represents about 60 percent of these costs. If one were to compare the
Army component only and assume Outrider is procured as an Army-only system, buying Hunter instead of Outrider
would save about $400 million in total costs.

c.   The Army plans to use Predators bought and operated by the Air Force, so they will cost the Army nothing.

d.   The costs of the Army's plan for Option III are based on the full Comanche program of 1,292 helicopters, not just the
number used in cavalry troops.

and support costs should be treated with considerable caution, however. Those costs are difficult to estimate for systems that have not yet finished their development and that the services have not had much experience with.

Option I: Cancel Outrider or Make It Solely an Army System

The first option focuses on DoD's highest priority for unmanned aerial vehicles: giving the Army's brigade commanders a UAV capability. The option would accomplish that in either of two ways, both of which are alternatives to trying to fulfill Army, Navy, and Marine Corps UAV requirements with Outrider. Outrider has suffered a number of technical problems during its development process, including excess weight, inability to meet the Navy's requirements for take-off and landing distances, and a delay in the development of a diesel engine for the air vehicle until after the ACTD. Furthermore, by design Outrider will not be a vertical take-off and landing UAV, which would be far more suitable for shipboard operations. Consequently, both alternatives under Option I would separate the Army's UAV requirements from those of the Navy and Marine Corps.

Option IA would cancel the Outrider program. In lieu of that system, the Army would use Hunter—which was developed in the mid-1990s and terminated after the production of 56 air vehicles—to fulfill its brigade-level UAV requirements. For their part, the Navy and Marine Corps would buy a UAV with vertical take-off and landing capability and a heavy-fuel engine. (The Navy has been looking at several such systems.) This option would save about $80 million in acquisition costs compared with buying Outrider for all three services, but it would increase costs by a total of about $10 million when 15-year operating and support costs are included.

The primary advantages of Option IA are that it would give Army brigades a more capable UAV system in a shorter amount of time than the Outrider program would, and it would give the Navy and Marine Corps a UAV system better suited to coastal warfare. The disadvantages are that Hunter requires substantially more transport aircraft than Outrider to deploy, and the replacements purchased for UAVs lost through attrition will probably be more expensive with Hunter than with Outrider because Hunter has a larger and more capable air vehicle.

Option IB attempts to address the same problems as Option IA but in a different way: by favoring the Army at the expense of the Navy and Marine Corps. Despite its problems, Outrider appears capable of meeting the Army's brigade-level requirements; thus, Option IB would buy that system solely for the Army. The Navy and Marine Corps would continue to rely on Pioneer for their UAV requirements. This option would save around $140 million in acquisition and operating and support costs compared with buying Outrider for all three services. The savings stem mainly from not buying a replacement for Pioneer.

The advantage of Option IB is that the Army would get the UAV capability that it clearly wants. The disadvantage is that the Navy and Marines would have to continue relying on an old UAV system that requires a great deal of maintenance.

Option II: Use Hunter to Meet the Army's Division and Corps UAV Requirements

Option II is designed to address the problems that might arise if the Army relies on Predators controlled by the Air Force to meet its division and corps UAV requirements. After Hunter was terminated in January 1996, the Army was left without a system to carry out its division and corps UAV missions. The Army proposes relying on the Air Force's Predator. But the Air Force plans to buy only 12 Predator systems, and about half that number would probably deploy in the event of a regional conflict. The Air Force has stated that although it is willing to use Predator to support division and corps commanders, higher priorities could be set by the theater commander or the national command authority that could require most, if not all, of the Predator assets. If the Army sent two corps and seven divisions to a regional conflict—as it did in the Gulf War—it seems unlikely that the average division commander would get a prompt response to his request for a Predator to perform a reconnaissance mission. One possible solution to that problem is to give each division and corps its own UAV capability using the Hunter systems the Army has in storage.

Option II would provide a Hunter system of four air vehicles, two ground control stations, and support equipment to every division and a system of six air vehicles, three ground control stations, and support equipment to every corps. In addition, the Air Force would continue to procure Predator for theaterwide use. This option would cost $250 million more for acquisition than the Army's plan to rely on Air Force Predators (which would cost it nothing). Including operating and support costs for 15 years, the option's price tag would total $750 million.

The principal advantage of Option II is that the Army's corps and divisions would get their own UAV systems. The disadvantages are the cost and the additional logistics required to deploy and maintain those Hunter systems in the field. The Air Force's Predators would deploy to a regional conflict regardless of how the Army plans to fulfill its corps and division requirements. Thus, the logistics involved in getting the Hunter systems to a theater would represent an additional burden over what the Army would require today.

Option III: Buy Tilt-Rotor UAVs and Reduce the
Army's Planned Comanche Helicopter Force

Could the Army benefit by deploying even more UAVs than it now plans? That is a difficult question to answer, but the Army Vice Chief of Staff did describe UAVs as a "major combat multiplier" after some exercises in which they played a prominent role. If more UAVs are deployed in the Army force structure, should they come at the expense of other assets, such as reconnaissance helicopters, or should they be in addition to them? As part of the development process for the Comanche reconnaissance helicopter, the Army was directed to analyze the "trade-offs" between the Comanche and unmanned aerial vehicles. While the Army is studying that issue, CBO has developed an option that would substitute tilt-rotor UAVs for many of the helicopters in the Army's cavalry aviation units. This option would save around $3.8 billion in acquisition costs and $700 million in 15-year operating and support costs.

The principal drawback of Option III is that the UAVs substituted into Army units in place of Comanches would not be armed. Thus, this option would sacrifice substantial combat capability. Aside from the money it would save, the advantage of this option is that in some ways the tilt-rotor UAVs are more capable reconnaissance platforms than the Comanche helicopters. They are about 15 percent faster and can watch an area five times longer before needing refueling. However, their line-of-sight communications link limits their radius of action to about 200 kilometers, whereas Comanches can travel much farther. However, UAVs are useful for more hazardous missions because they do not risk the lives of an aircrew.

Option IV: Use Global Hawk UAVs to Substitute for the Reduction of JSTARS

Option IV would provide additional Global Hawk UAVs to supplement the Joint Surveillance Target Attack Radar System (JSTARS) fleet. JSTARS is a joint Army/Air Force reconnaissance system that combines a powerful multimode ground-surveillance radar with command-and-control systems on board a 707 aircraft. The purpose of JSTARS is to detect mobile and stationary targets on the ground and transmit their locations to ground commanders and combat aircraft. DoD had planned to buy 19 such systems to provide continuous coverage of two theaters of combat simultaneously. The recent Quadrennial Defense Review, however, proposed reducing that planned purchase to 13 aircraft (plus one for testing).

In the Quadrennial Defense Review report, DoD argued that a fleet of 13 JSTARS aircraft would be able provide the round-the-clock coverage needed in a major theater war. In the event of a second war, some of the aircraft would have to be redeployed to the second theater, possibly opening gaps in coverage. The Department of Defense plans to "explore the potential for supplementing radar coverage of enemy force movements from long-endurance unmanned aerial vehicles."[1] CBO's Option IV reflects that idea.

This option would supplement the reconnaissance capability of the reduced JSTARS fleet by buying some additional Global Hawks to support the JSTARS mission. The option would cost almost $500 million more in acquisition costs and $700 million in 15-year operating and support costs than the Air Force plans to spend on either the Global Hawk or JSTARS program. But in return for that additional cost the Air Force would get additional capability. In particular, because Global Hawk would not put an aircrew in jeopardy, it could be deployed far deeper into enemy territory than the JSTARS aircraft.

Option V: End Darkstar After the ACTD and Rely on Other Systems

CBO's last option would end production of Darkstar with the three air vehicles left over from the Advanced Concept Technology Demonstration. It seeks to address Congressional concerns about apparent overlaps in the unmanned aerial vehicle programs. Darkstar is a high-altitude UAV that is designed to have low-observable (stealthy) characteristics. It is intended to carry out a particular mission: collecting imagery over highly defended targets before an enemy's air defenses have been suppressed. In addition, because of its stealthy characteristics, it is likely to be useful in supporting special-operations forces.

Other than stealth, Darkstar is expected to be a less capable UAV than Global Hawk but more capable (except for endurance) than Predator (see Summary Table 2). The Defense Airborne Reconnaissance Office and the Air Force have described Global Hawk as a highly capable but moderately survivable UAV, whereas Darkstar is a highly survivable but moderately capable UAV. The chief advantage of buying Darkstar, therefore, is to buy stealthy reconnaissance capability.

Option V would save $600 million in acquisition costs and another $400 million in 15 years of operation and support. In a sense, that is the price DoD and the Air Force appear willing to pay for stealth in an unmanned aerial vehicle. The advantage of this option is that it would save money. The disadvantage is that the Air Force would have only a limited stealthy UAV capability (just three air vehicles). However, other UAVs, such as Global Hawk and Predator, may be able to perform many of Darkstar's missions, albeit at a greater risk of being shot down by enemy air defenses. Furthermore, in light of the less threatening environment that the United

---

1.   Secretary of Defense William S. Cohen, *Report of the Quadrennial Defense Review* (May 1997), p. 45.

States faces today compared with during the Cold War, ending the Darkstar program may be an acceptable risk to take.

CHAPTER I

PROGRAMS AND MISSIONS FOR

UNMANNED AERIAL VEHICLES

Many defense analysts argue that the nature of warfare and the way the United States will fight future wars are undergoing a fundamental transformation. They contend that the development of new technologies (such as stealthy aircraft, highly accurate precision munitions, and improved sensors for detecting, tracking, and identifying enemy forces) will work together to allow a force to dominate the battlefield completely—more so than has been achieved in the past, even during the Persian Gulf War. In short, that revolution in military affairs means having a monopoly on information about the battlefield, as well as the ability to attack and destroy an enemy while denying it the same capability.

The armed services have long had many different ways to collect battlefield intelligence. The scout on foot is probably the earliest example. Today, the U.S. military also uses sensors that are mounted on a variety of satellites, manned aircraft, helicopters, and ground vehicles to collect information. In the future, it also hopes to make greater use of unmanned aerial vehicles (UAVs) to carry sensors.

UAVs, which have sometimes been referred to as drones, are relatively small aircraft that can be preprogrammed or operated by remote control. Many defense analysts view them as crucial to the success of the revolution in military affairs. In the course of one mission, a UAV can find, identify, and even direct a precision munition to a target—and then assess the damage done to that target after the munition has hit—without risking the lives of an aircrew. UAVs are also appealing to the military because different UAV systems can collect different types of information, such as tactical (or battlefield) intelligence and strategic (or longer-range) intelligence. In addition, UAVs may be able to perform such roles as relaying messages during a battle, locating or jamming enemy radar, or monitoring areas during peacekeeping missions.

The services are developing and plan to procure four new UAV systems: Predator, Darkstar, Global Hawk, and Outrider (see Table 1). Those programs, if ultimately successfully, promise to give battlefield commanders a valuable new reconnaissance capability as well as to enhance and perhaps eventually replace many sophisticated manned reconnaissance systems that provide intelligence to theater commanders and the national command authority (the President and the Secretary of Defense).

Although the Congress generally supports UAV technology, it has expressed concern at the proliferation of UAV programs—particularly in light of their troubled technological history and the seeming inability of the Department of Defense (DoD) to develop and field a major UAV system. In analyzing DoD's current programs, the Congressional Budget Office (CBO) has identified three key areas of concern. They are the suitability of some new UAVs for their intended missions, apparent overlaps in capability among different systems, and uncertainty about who will control UAVs on the battlefield.

TABLE 1.   MAJOR UNMANNED AERIAL VEHICLE PROGRAMS

Program	Period	Description	Status
Lightning Bug	1964-1979	Reconnaissance drone first used by the Air Force during the Vietnam War	Retired
Aquila	1979-1987	Tactical UAV for Army commanders	Canceled
Amber	1984-1990	Classified endurance UAV	Canceled
Pioneer	1986-present	UAV originally acquired to assess battle damage by naval gunfire	Deployed
Medium Range	1987-1993	Tactical UAV for the Air Force and Navy	Canceled
Hunter	1988-1996	Joint tactical UAV	Canceled after LRIP[a]
Gnat-750	1988-present	Long-endurance UAV developed with CIA funding; exported commercially	Used for training and intelligence missions
Predator	1994-present	Long-endurance UAV for theater commanders; based on the Gnat-750	In LRIP
Darkstar	1994-present	Stealthy endurance UAV for high-threat environments	In development
Global Hawk	1994-present	High-altitude, long-range endurance UAV	In development
Outrider	1996-present	Joint tactical UAV	In development

SOURCE: Congressional Budget Office.

NOTE: UAV = unmanned aerial vehicle; LRIP = low-rate initial production; CIA = Central Intelligence Agency.

a.   Seven systems—each equipped with eight air vehicles, four ground control stations, and support equipment—were initially placed in storage. Later, most of the equipment for two systems was removed and used in training exercises and in developing "concepts of operation" for UAVs.

A BRIEF HISTORY OF UAVs

UAVs, in one form or another, have had a checkered history in the U.S. military. Although the notion of using unmanned aircraft has been around since World War I, the United States did not begin seriously experimenting with unmanned reconnaissance drones until the late 1950s. That initial effort proved unsuccessful. Later, the Vietnam War and the Cold War spurred a variety of development programs, which led to several reconnaissance drones, such as the Firefly and Lightning Bug. Although those early UAVs were sometimes difficult to operate and maintain, the Air Force deployed them for a variety of missions, including gathering signals intelligence and collecting high- and low-altitude imagery both during the day and at night. By the end of the Vietnam War, concern about casualties meant that only two aircraft were allowed to fly reconnaissance missions over North Vietnam: the Lightning Bug UAV and a high-altitude, manned reconnaissance plane (the supersonic SR-71).

The urgent need for unmanned aerial vehicles ended with the Vietnam War, but the services remained interested in exploring the capabilities that those aircraft had to offer. In particular, the Army began developing a tactical UAV called Aquila in 1979. It suffered many growing pains (developmental problems, cost overruns, changes in requirements) and was finally canceled in 1987. During that time, the Israelis used very simple and cheap drones to good effect to destroy Syrian air defenses in Lebanon's Bekaa Valley in 1982. Their success inspired then Secretary of the Navy John Lehman to push for his service to acquire UAVs, primarily to support targeting by, and conduct battle-damage assessment for, U.S. battleships. His efforts led the Navy and Marine Corps to acquire nine Pioneer UAV systems, which are still in use today. Those systems have been employed in many U.S. operations since the 1980s, including the Gulf War and Bosnia. In addition, the armed forces, particularly the Marine Corps, have used some very small UAVs, such as the Exdrone, in both operations and training.

In recent years, the Pentagon has started a number of other UAV development programs. Two of them—Medium Range and Hunter—were ultimately canceled. Another UAV, Predator, is now being acquired by the Air Force. And three others—Darkstar, Global Hawk, and Outrider—are still in development. DoD officials appear more optimistic about this group of UAVs than about earlier ones, partly because advances in technologies such as miniaturization make developing UAVs easier, and partly because the developers now have more experience in integrating all of the components that compose a UAV system (such as the air vehicle, ground support equipment, sensors or other payloads, and communications equipment).

THE ROLE OF UAVs IN FUTURE WARFARE

When DoD or the services attempt to envision the future of warfare, UAVs play an important role in their vision. One of the central concepts in predictions about future warfare is the so-called revolution in military affairs. That term describes a group of technologies (long-range precision munitions; stealthy aircraft "platforms"; real-time, all-weather, day-and-night reconnaissance and targeting; and integration of command and control among the services) that, once combined, produce a major leap in a unit's fighting power. The revolution in military affairs also includes innovations in strategy, operations, and tactics, which in turn are reflected in the training programs of the services. For example, one priority of that revolution is being able to mass fire from widely dispersed forces to have a concentrated effect on one location. Achieving that requires having improved reconnaissance, communications, and precision-strike capabilities, as well as new tactics that must be incorporated into training programs.

The official vision statement of the Joint Chiefs of Staff, *Joint Vision 2010*, fully embraces the revolution in military affairs. A crucial component of *Joint Vision 2010* is the importance of information superiority—the ability to collect, process, and disseminate an uninterrupted flow of information while exploiting or denying an enemy's ability to do the same. UAVs are likely to be crucial in achieving information superiority, particularly because they can collect information that in the past would have been difficult to acquire without risking the lives of personnel. Although the text of *Joint Vision 2010* is not specific about which weapon systems and platforms should form the future force, the graphics that accompany the report give a prominent place to unmanned aerial vehicles.

DoD's Integrated Airborne Reconnaissance Strategy also recognizes and incorporates UAVs. That strategy examines the future reconnaissance needs of the services, the technologies and platforms necessary to meet those needs, and ways to integrate those technologies so they are more cost-effective and can be operated by a variety of DoD's warfighting elements. Over the long run, the strategy expects UAVs to provide wide-area surveillance and continuous coverage at the strategic level and possibly replace manned tactical reconnaissance altogether. In addition, unmanned aerial vehicles are expected to give national decisionmakers greater willingness to accept the risks normally associated with airborne reconnaissance. That does not necessarily mean that UAVs will survive better in high-threat environments than manned systems, but that they can be risked without endangering the lives of aircrews.

Vision statements or studies by the various services also include an important role for unmanned aerial vehicles. For example, the Navy's littoral (coastal) warfare strategy, *Forward . . . From the Sea,* and the Marines Corps's concept of operational maneuver from the sea incorporate UAVs to provide timely reconnaissance on the prospective enemy forces that an amphibious task force may confront. They suggest that UAVs may also be able to provide targeting information to long-range precision weapons and then follow up with battle-damage assessments of those targets. Similarly, studies by the Air Force's Scientific Advisory Board and DoD's Defense Science Board have envisioned a central role for UAVs in future combat operations.

The Potential Missions of Unmanned Aerial Vehicles

Defense planners expect UAVs to perform a variety of specific missions, many of which fall into the broad category of reconnaissance and surveillance. Those missions exclude the more advanced concepts that are sometimes discussed of using UAVs directly as combat vehicles.[1] Such ideas are just beginning to be developed and are beyond the scope of this analysis.

The missions that UAVs can perform fall into two general categories of intelligence: tactical and strategic. Tactical intelligence refers to information collected for the local area of operations. For example, a brigade commander whose unit is responsible for seizing the ground in his area of operations will want to know the size, quality, and disposition of any enemy forces in that area. In the past, a brigade commander usually had to rely on troops or manned aircraft, such as helicopters, to scout that terrain and relay the information back to him. But a UAV designed to support such a commander could provide him with nearly instantaneous (or near-real-time) video, day or night, of the terrain "just over the hill" without risking pilots or scouts. That, the Army stresses, is a capability brigade commanders have never had before. Furthermore, the tactical UAVs now under development should be able to provide surveillance for a much longer time before requiring relief than a manned platform can.

A good illustration of that capability was provided during the Army's Task Force XXI Advanced Warfighting Experiment at the National Training Center in Fort Irwin, California. The Army praised the tactical UAV involved in the war game (Hunter) for greatly aiding the commander of the task force. According to the Army, Hunter's presence caused the opposing commander to spend an extraordinary amount of time protecting his own force and thus made it difficult for him to assemble that force for an attack.[2]

---

1.   Such combat-related missions may one day include using UAVs to conduct strike missions, suppress enemy air defenses, or carry nonlethal weapons, such as a high-powered microwave that could disrupt an enemy's electronic equipment.

2.   Ltc. William L. Burnham, "TF Hunter Support to AWE" (briefing by the Army, National Training Center, March 1997).

In contrast to tactical intelligence, strategic intelligence refers to the type of longer-range information collected by reconnaissance satellites and manned U-2 aircraft. It can include information about another country's military assets, such as the concentration of its forces, weapons of mass destruction, and industrial and manufacturing facilities. Some of the UAVs under development are designed to provide that type of information. The most capable ones could monitor a particular area for more than 24 hours at a time—longer than an orbiting satellite can and without the risks of manned reconnaissance flights.

Reconnaissance and Surveillance. Although some people use the terms "reconnaissance" and "surveillance" interchangeably, this analysis makes a distinction between the two. Reconnaissance refers to obtaining information about the activities or resources of an enemy (or potential enemy) or collecting data about the meteorological, hydrographic, or geographic characteristics of a particular area. Unmanned aerial vehicles can perform that mission, although the size of the area they can cover will be different for different types of UAVs. For example, a wide-area search collector, such as Global Hawk, will be able to cover dramatically greater expanses than a small tactical UAV, but both are capable of searching for enemy forces.

Surveillance, by contrast, refers to watching a particular site or road or target for an extended period of time. For example, other sources of intelligence may identify a building as a possible hiding place for weapons (such as mobile ballistic missiles) but may not be certain that that is the case. A UAV, particularly one with a relatively long operating period (or "endurance"), can watch the building and observe whether mobile missiles arrive or depart. Or it may confirm, because of other activity, that the building is not a storage facility. According to DoD officials, the Air Force's Predator UAV has been used several times to follow a particular vehicle along a road for an extended period. At the end of the trip, the vehicle—in this case, a military truck—entered a building that was not known to be a facility for housing or hiding weapons. Through the imagery provided by the UAV, intelligence officers were able to determine that it was such a facility.

Another important surveillance role that unmanned aerial vehicles can play is in peacekeeping operations. For example, if a neutral zone could not be patrolled by peacekeepers on the ground (perhaps because of difficult terrain), a long-endurance UAV could watch the zone for violations. If several UAVs were available, the commander of a peacekeeping operation could conceivably have continuous video coverage of a disputed, demilitarized area. Predator successfully performed some similar functions during peacekeeping operations in Bosnia.

Target Acquisition. A principal mission of unmanned aerial vehicles is target acquisition—that is, detecting, locating, and identifying a particular target. A wide-area search aircraft, such as the Joint Surveillance Target Attack Radar System (JSTARS), might identify a group of moving vehicles in a particular area. A UAV could then be directed to that location to confirm, for example, whether the vehicles were tanks and, if so, whose they were. The UAV could also relay fairly precise location information so the vehicles could be attacked.

Target Designation. Some tactical unmanned aerial vehicles have both the power and payload capacity to carry a laser target designator. That device aims a laser at a target, such as a tank, so another platform can attack it with a precision munition. The Army has conducted at least four successful tests in which a Hunter UAV mounted with a laser designated a target, and another aircraft, such as a Kiowa Warrior helicopter, launched a Hellfire missile and destroyed the target.

Communications Relay. An unmanned aerial vehicle can also serve as a communications relay platform. In battle, forces may move quickly and exceed the range of their communications system, as happened during the Gulf War.[3] A communications relay carried on board a UAV might be able to bridge the gap between the leading edge of a U.S. advance and higher-echelon commanders located farther back. Similarly, even in cases in which communications systems use satellites to give them longer range, UAVs could perform the same relay role as a satellite if the latter became overwhelmed with traffic or was damaged. Global Hawk has a far greater range, payload, and speed than any other planned UAV, so it probably has the most potential to substitute for or supplement satellites.

Battle-Damage Assessment. The original mission for which the Navy acquired UAVs in the 1980s was to assess the damage inflicted by its weapons. Specifically, the Navy bought Pioneer in part because it determined that long-range fire from battleships could be much more effective and precise if a UAV was watching the target area. The UAV substituted for a human forward observer to give gunnery crews direct imagery about how close a salvo came to hitting the target and what corrections were needed to strike precisely. Even though battleships have been retired, UAVs can perform the same mission over targets being attacked by aircraft, missiles, or other ships.

Communications and Electronics Intelligence. The Department of Defense conducted tests of a communications intelligence payload on board a Hunter UAV in November 1996. The primary purpose of such a payload is to locate and identify an enemy force's ground communications emitters, such as radio transmitters. In some cases, that can lead to detecting and identifying the locations of the enemy's military leaders as they communicate with their forces.

---

3.   Michael R. Gordon and General Bernard E. Trainor, *The Generals' War: The Inside Story of the Conflict in the Gulf* (Boston: Little Brown, 1995), pp. 387-389.

At the same time, DoD also tested an electronics intelligence payload on board Hunter that was designed to find and identify an enemy's ground radar emitters. Doing that successfully can enable U.S. forces to attack and destroy those installations early in a conflict, bringing serious harm to the enemy's air-defense network.[4]

Jamming. Besides locating enemy communications or radars, UAVs could be used to jam them. DoD has also tested payloads on Hunter that have employed a radar jammer or communications jammer.

Chemical and Biological Warfare Detection. Unmanned aerial vehicles are well suited to detect and determine the lethality of environments that are contaminated by chemical or biological agents. Using specialized payloads, UAVs can sample different layers of air over an area suspected of contamination without risking the lives of soldiers.

How Would UAVs Affect a Battle?

Once the currently planned reconnaissance UAVs are finished and deployed with U.S. forces, what difference will they make? The most obvious answer is that their widespread use will almost certainly save lives. If nothing else, risky reconnaissance missions that in the past were performed by manned aircraft or scout units could now be performed by unmanned aerial vehicles. Those missions include normal reconnaissance operations that are at risk from enemy fire as well as missions in environments contaminated by nuclear, chemical, or biological weapons.

Beyond that, how much difference UAVs will make to the success of a battle varies with the capabilities and response of the opponent. For example, few unmanned aerial vehicles were used during the Gulf War. More widespread use in that conflict would probably have made very little difference to the outcome because the war was so one-sided—both because of U.S. skill, strategy, and weapons and because of mistakes by the Iraqi army.

More insight can be gained from the Army's Advanced Warfighting Experiment at the National Training Center. In those exercises, the training units, which were equipped with UAVs, gained an advantage over the permanently based opposition force (OPFOR), which usually wins the "battles" at the training center. The commanders of both the opposition force and the units training there attributed a high military value to UAVs. According to some Army officials, the opposition-

---

4.   See, for example, David A. Fulghum, "UAV Succeeds in Electronic Combat," *Aviation Week & Space Technology,* January 26, 1998, p. 29.

force commander became so concerned about destroying the UAVs that he changed his tactics and thus created opportunities for other systems to be more effective than they would otherwise have been. If those exercises are any guide, when unmanned aerial vehicles are used against a force that is not similarly equipped, they can enable military commanders to observe the enemy, identify its location and targets, and thereby disrupt its strategy—whether offensive or defensive—before that strategy can be carried out.

A more difficult question to answer—and one that has been asked in Congressional hearings—is, what would happen if both the United States and its opponent in some future conflict were similarly equipped with unmanned aerial vehicles? Assuming that both sides' commanders were intelligent and capable and both had UAVs, the presence of those vehicles would make little difference. The outcome would again depend on which side had the better-led, better-equipped forces—just as if both sides did not have UAVs. At the National Training Center, opposition-force commanders have asked for their own UAVs in future exercises to balance the battlefield. Exercises in which both sides operate unmanned aerial vehicles may provide some answers to that question.

DoD'S UAV PROGRAMS

As a practical matter, the Pentagon has decided that one type of UAV is not enough to provide imagery at all of the necessary levels. For example, the reconnaissance needs of a brigade commander, who controls a few thousand soldiers, are very different from those of the theater commander, who is in charge of the entire area of operations. The former needs to know what is a few kilometers away (or even just over the next hill), and television video is probably the most useful imagery to convey that information. The theater commander is concerned with a much larger number of issues, over a much wider area, and thus requires a much more capable and flexible platform for collecting intelligence.

Because of those different needs, the Department of Defense is developing a family of unmanned aerial vehicles to meet the multiple imagery requirements of the various services. One UAV (Pioneer) is already deployed; another (Hunter) was canceled after initial production, but the remaining systems are used in some training exercises. Predator has moved from the development stage to low-rate initial production for the Air Force. Three other UAVs (Darkstar, Global Hawk, and Outrider) are still being developed.

Pioneer

Pioneer has provided the Navy and Marine Corps with UAV capability since 1986.[5] It was first acquired to support targeting by the 16-inch guns on board battleships that the Navy had brought back into service. After those ships were retired, the Navy and Marines kept the Pioneer systems to provide near-real-time reconnaissance, surveillance, target acquisition, battle-damage assessment, and battle management in both day and night operations. Pioneer can carry a 75-pound payload and provide five hours of endurance at a range of 185 kilometers (see Table 2). It uses a line-of-sight communications and data link, meaning that the UAV cannot operate over the horizon and still communicate with its controllers.

Both the Navy and Marine Corps consider Pioneer an outmoded system, however, and are anticipating its eventual replacement by either the Outrider tactical UAV or some other, as yet undetermined, system. In the meantime, Pioneer's service life has been extended several times. Currently, the last Pioneer is expected to retire in 2003, but the Navy and Marine Corps are considering retaining Pioneers until 2005 or 2008—probably as a result of the development problems that Outrider is experiencing and the Navy's desire to have a UAV with a heavy-fuel engine and a vertical take-off and landing capability.

While awaiting a successor, the Navy and DoD are making a number of upgrades to Pioneer to extend its service life to at least 2003. One involves integrating Pioneer with the common automatic recovery system, a technology (both hardware and software) that will allow the air vehicle to land automatically, thus increasing safety and reducing mishaps. Other enhancements include improving its ability to tap into the Global Positioning System, integrating Pioneer with the tactical control system (itself under development) that is designed to enhance the interoperability of UAV systems and their controllers, and incorporating a common tactical data link for the same purpose. Further upgrades to improve Pioneer's reliability are being considered. Modifications to enhance the capability of the air vehicle—such as a redesigned wing to double its endurance—are also possible, but they are currently restricted by a DoD policy that permits only reliability upgrades.

Hunter

The Hunter unmanned aerial vehicle grew out of the operational requirements document for the Short Range UAV system that DoD published in 1992. That document specified a military requirement for a tactical UAV with a radius of about

---

5.   The Army also used Pioneer but gave it up in 1995 in anticipation of the initial operational capability of Hunter.

TABLE 2.   OPERATIONAL FACTORS FOR DEPLOYED UAVs

	Pioneer	Hunter	Predator
Radius (Kilometers)[a]	185	267	926
Endurance at Radius (Hours)	5	11	20 or more
Total Endurance (Hours)	7	14	35
Typical Operating Altitude (Feet)	3,000-8,000	10,000	10,000-25,000
Maximum Altitude (Feet)	15,000	15,000	25,000
Cruise Speed (Kilometers per hour)[b]	120	165	120
Dash Speed (Kilometers per hour)[c]	175	196	130
Types of Sensors	EO or IR	EO and IR	SAR, EO, and IR
Payload (Pounds)	75	200	450

SOURCE: Congressional Budget Office based on data from the Department of Defense, the Army, and the Air Force.

NOTE: UAV = unmanned aerial vehicle; EO = electro-optical (video); IR = infrared; SAR = synthetic aperture radar.

a.   Expected operating range.

b.   Normal operating speed.

c.   Maximum speed.

200 kilometers and endurance of eight to 12 hours that would provide imagery to commanders of corps, divisions, and task forces. Hunter was specifically designed to meet those requirements. It has a radius of 267 kilometers and an endurance at that radius of about 11 hours. It can carry roughly 200 pounds and provide imagery in both day and night operations.

Problems in Hunter's development, however, led to its early demise. Three crashes in 45 days, as well as its inability to meet some of its performance criteria, led the Joint Requirements Oversight Council to recommend canceling the program in October 1995.[6] The General Accounting Office also issued reports critical of

---

6.   The Joint Requirements Oversight Council is a body that reviews defense acquisition programs. It is composed of the vice chiefs of staff of the services and chaired by the Vice Chairman of the Joint Chiefs of Staff.

Hunter and ultimately recommended its cancellation.[7] The contract for low-rate initial production expired on January 31, 1996, and the acquisition program was terminated. Seven systems, each with eight air vehicles, were produced before cancellation and initially placed in storage.

Since then, however, Hunter has managed an impressive resurrection. The equipment of almost two complete systems has been removed from storage and is being used by the Army and Navy in training programs and in developing "concepts of operation" for how and when UAVs should be used by soldiers in the field. As part of those efforts, Hunters were employed in the Army's Advanced Warfighting Experiment. All seven Hunter systems have received a host of reliability upgrades, including improvements to the engine, data link, and flight control. In 1996 and 1997, Hunter flew over 3,000 hours of operations with only two serious mishaps. Indeed, an official with the Unmanned Aerial Vehicles Joint Program Office described Hunter as "the most reliable UAV program we have."[8]

The services, especially the Army, have also used Hunter to demonstrate various missions that UAVs are capable of performing, such as carrying communications and electronics intelligence payloads and laser target designators. In addition, the Navy has used parts of a Hunter system to practice integrating a UAV system more closely with Navy units.

Predator

Aside from Pioneer, Predator is the only UAV to have moved from development to full acquisition to operational deployment. It grew out of the Gnat-750 and Amber programs developed in the 1980s for the Defense Advanced Research Projects Agency and the Central Intelligence Agency.[9] In January 1994, DoD gave General Atomics a contract to develop Predator under the Advanced Concept Technology Demonstration (ACTD) process, which is intended to give the eventual users of a system more role in its development and testing. Predator's first flight took place six

---

7.   General Accounting Office, *Unmanned Aerial Vehicles: No More Hunter Systems Should Be Bought Until Problems Are Fixed*, NSIAD-95-52 (March 1, 1995); and General Accounting Office, *Unmanned Aerial Vehicles: Hunter System Is Not Appropriate for Navy Fleet Use*, NSIAD-96-2 (December 1, 1995).

8.   "Official says 'Hunter most reliable UAV'; additional buy sought," *Aerospace Daily*, November 18, 1997, p. 267A.

9.   Curtis Peebles, *Dark Eagles: A History of Top Secret U.S. Aircraft Programs* (Novato, Calif.: Presidio, 1995), pp. 209-214. See also Michael R. Thirtle, Robert U. Johnson, and John L. Birkler, *The Predator ACTD: A Case Study for Transition Planning and the Formal Acquisition Process* (Santa Monica, Calif.: RAND, 1997), p. 8.

months later, and the ACTD process was completed 18 months later with a decision to proceed to low-rate acquisition of the system in July 1996. That is not to say that the development process was entirely smooth or without problems. (For more on the ACTD process and how it affected Predator's development, see Chapter II.)

Predator is a medium-altitude, medium-range UAV that can provide near-real-time reconnaissance, surveillance, target acquisition, and battle-damage assessment day or night and in some difficult weather conditions.[10] Predator's normal operating altitude is 15,000 feet, although it can function as high as 25,000 feet. Its maximum speed, 130 kilometers per hour, is actually slower than that of other existing and planned tactical UAV systems; its normal operating (or "cruise") speed is about 120 kilometers per hour. Aside from the sensor payload that the air vehicle carries, Predator's real assets are its endurance—more than 20 hours at its radius of 926 kilometers—and its communications system, which includes a satellite link. That means the air vehicle can operate beyond the line of sight of the ground control station and still relay images back to the user. Currently, the Air Force expects to acquire 12 Predator systems, each with four air vehicles and one ground control station.

The principal mission of Predator is to support the reconnaissance needs of the theater commander in a regional conflict. As such, it is not a system likely to be controlled by lower-echelon officers such as brigade commanders. After the cancellation of Hunter, however, the Army expressed hope that its division and corps commanders could use Predator to provide imagery in support of their missions. But unless the Army buys its own Predator systems, its division and corps commanders seem unlikely to get much support from the UAV, partly because so few systems are being bought and because a regional conflict would probably involve many higher-priority missions.

Predator has demonstrated its military usefulness in peacekeeping missions such as the recent ones in Bosnia. During the ACTD stage, several Predators were sent to Bosnia, where they helped NATO military commanders enforce the terms of the cease-fire. That included detecting troop movements in unauthorized areas, discovering previously unknown weapons factories or depots, and locating units that were breaking the peace.

---

10.   The limitations that Predator does have in bad weather do not result from its sensor—which includes a synthetic aperture radar that can provide imagery through clouds and rain—but from the air vehicle itself. High winds and severe storms can sometimes impair or prevent the air vehicle from operating or from successfully executing its mission.

Darkstar and Global Hawk

Darkstar and Global Hawk are high-altitude unmanned aerial vehicles that will provide reconnaissance, surveillance, and target-acquisition information to theater and higher-echelon commanders. Both UAVs are being developed as a single ACTD by the Defense Advanced Research Projects Agency, in part because the missions they are expected to perform complement each other. Global Hawk (also known as Tier II Plus) is supposed to be a "highly capable, moderately survivable" UAV, whereas Darkstar (also known as Tier III Minus) is expected to be a "moderately capable, highly survivable" UAV. The developers expect Global Hawk to have a maximum operating altitude of 65,000 feet and a radius of 5,556 kilometers, with about 24 hours of endurance at that distance (see Table 3). Darkstar is expected to have a maximum operating altitude of 45,000 feet and a radius of 926 kilometers, with around eight hours of endurance at that radius.

Global Hawk is a large aircraft—almost as big as a U-2 manned recon-naissance plane (see Table 4). It was not designed to be stealthy and therefore is likely to be vulnerable to high-altitude surface-to-air missiles. In contrast, Darkstar is being developed as a low-observable (stealthy) UAV that can penetrate enemy air defenses, perform its mission, and return. According to DoD, combining the capabilities of Global Hawk and Darkstar into one aircraft would be both technically difficult and extraordinarily expensive—hence the decision to pursue two separate air vehicles.

Both Global Hawk and Darkstar are being developed with a philosophy that uses cost as an independent variable. The only firm requirement is that the average cost of the 11th through 20th air vehicles—for both types of UAV—be no more than $10 million (in 1994 dollars). All other technical characteristics can be traded to fulfill that requirement.

The technical objectives for the two high-altitude UAVs stem from some broadly worded statements of mission needs for reconnaissance, surveillance, and target acquisition. Essentially, the requirement is for systems that can provide near-real-time reconnaissance, surveillance, and target acquisition to theater, midlevel, and tactical commanders and that can operate in a variety of environments—defended and undefended, contaminated and uncontaminated—without risking the lives of soldiers. For example, in justifying the development and procurement of both manned and unmanned long-range reconnaissance systems, the Joint Requirements Oversight Council determined that:

TABLE 3.   TECHNICAL OBJECTIVES FOR UAVs UNDER DEVELOPMENT

	Darkstar	Global Hawk	Outrider
Radius (Kilometers)[a]	926	5,556	200
Endurance at Radius (Hours)	8	22	3-4
Total Endurance (Hours)	12	40	6
Typical Operating Altitude (Feet)	40,000-45,000	55,000-65,000	5,000-10,000
Maximum Altitude (Feet)	45,000	65,000	15,000
Cruise Speed (Kilometers per hour)[b]	463	639	140
Dash Speed (Kilometers per hour)[c]	n.a.	n.a.	204
Types of Sensors	EO or SAR	SAR, EO, and IR	EO and IR
Payload (Pounds)	1,000	2,000	65

SOURCE: Congressional Budget Office based on data from the Department of Defense and the Air Force.

NOTE: UAV = unmanned aerial vehicle; n.a. = not available; EO = electro-optical (video); IR = infrared; SAR = synthetic aperture radar.

a.   Expected operating range.

b.   Normal operating speed.

c.   Maximum speed.

Warfighting commanders-in-chief (CINCs) have a need to provide commanders [with] a responsive capability to conduct wide-area near-real-time reconnaissance, surveillance, and target acquisition (RSTA), command and control, signals intelligence (SIGINT), electronic warfare (EW), and special operations missions during peacetime and all levels of war against defended/denied areas over extended periods of time. The evolution of the hostile surface-to-air and air-to-air threat and their collective effectiveness against manned aircraft and satellites can generate unacceptably high attrition rates. Current systems cannot perform these missions in a timely, responsive manner in an integrated hostile air defense environment without high risk to personnel and costly systems. There is a need for a capability which can be employed in areas where enemy air

TABLE 4.   COMPARISON OF THE SIZE OF UAVs AND MANNED AIRCRAFT

	In Feet				In Pounds
Aircraft	Wingspan	Length	Height	Width	Weight[a]
Unmanned Aerial Vehicles					
Outrider	13	10	5	1.3	n.a.
Pioneer	17	14	3	1.5	463
Hunter	29	23	6	n.a.	1,600
Predator	48	27	7	3.7	2,230
Darkstar	69	15	5	12.0	8,600
Global Hawk	116	44	15	6.0	25,600
Manned Aircraft					
F-16C/D	31	49	17	n.a.	42,300
U-2R	103	63	16	n.a.	40,000

SOURCE: Congressional Budget Office based on data from Aviation Week & Space Technology and the Association for
Unmanned Aerial Vehicle Systems International, *1997-1998 International Guide to Unmanned Aerial Vehicles*
(New York: McGraw Hill, 1997); John W.R. Taylor, ed., *Jane's All the World's Aircraft, 1989-1990* (Coulsdon,
Surrey, England: Janes's Information Group, 1989); and Paul Jackson, ed., *Jane's All the World's Aircraft,
1995-1996* (Coulsdon, Surrey, England: Jane's Information Group, 1995).

NOTE: n.a. = not available.

a.   Maximum take-off weight.

defenses have not been adequately suppressed, in heavily defended areas, in open ocean environments, and in contaminated environments. Nuclear survivability is required as necessary to perform missions in a nuclear contaminated environment, including operating in the presence of high-altitude EMP [electromagnetic pulses].[11]

---

11.   Department of Defense, Joint Requirements Oversight Council, *Mission Need Statement for a Long Endurance, Reconnaissance, Surveillance, and Target Acquisition Capability,* JROCM-003-90 (January 5, 1990).

The Global Hawk and Darkstar ACTD has had various problems. Both air vehicles have experienced substantial slips in their schedules. The first Darkstar aircraft crashed on its second flight. And there are reasons to believe that both aircraft will have difficulty meeting their cost goals, although that will not be known for sure until the end of the ACTD process.

Outrider

Outrider is a tactical UAV that is being developed for the joint use of the Army, Navy, and Marine Corps. It is intended primarily to meet the reconnaissance and surveillance needs of brigade and task-force commanders. Like Darkstar and Global Hawk, Outrider is currently under development as an ACTD, and its only firm requirement is that its cost be no more than $350,000 by the 33rd air vehicle and no more than $300,000 by the 100th. Other technical objectives, which were laid down by the Joint Requirements Oversight Council, may be traded off against one another to achieve that cost requirement.

The technical objectives for Outrider stem from DoD's operational requirements documents for the previously planned Close Range and Short Range UAV systems. That earlier approach to tactical UAVs envisioned having two systems: one (with a radius of about 50 kilometers and endurance of three to four hours) that would provide imagery to brigade-level commanders, and another, longer-range UAV that would fulfill the requirements of corps and division commanders. As noted above, Hunter was intended to be the second system, the Short Range UAV. After Hunter was canceled in January 1996, the program to develop a tactical UAV—which was still the first priority of the Joint Requirements Oversight Council—was converted to an ACTD, and the technical objectives were set so that the new system would fulfill missions of both the Army and the Navy and Marine Corps. The system selected was Alliant Techsystem's Outrider, which had originally been proposed as the Close Range UAV.

Developing Outrider to meet the joint requirements has proved challenging. The program has experienced numerous schedule slips, and the air vehicle has had trouble meeting some of its performance goals during tests. The troubles were severe enough that in September 1997 the General Accounting Office issued a report recommending a delay in production of Outrider until all of the problems have been solved and the system's military utility has been demonstrated.[12] However, officials of the Joint Program Office, which is in charge of managing the Outrider program, are cautiously optimistic that Outrider's problems will be overcome and that the

---

12.   General Accounting Office, *Unmanned Aerial Vehicles: Outrider Demonstrations Will Be Inadequate to Justify Further Production*, NSIAD-97-153 (September 1997).

ACTD will yield a useful system. Outrider has had more than 160 test flights, and program officials appear to be mitigating some of the technical problems, such as the excess weight of the air vehicle. Thus far, however, Outrider has had only a handful of test flights longer than three hours.

PROBLEMS WITH DoD'S UAV PROGRAMS

Although unmanned aerial vehicles appear to show great promise and many people have high expectations for them, the Congress is concerned that so many of the UAV systems that DoD has developed or is developing have experienced problems. Historically, many of the services' UAV programs have run into technical difficulties and cost growth. (For example, the Army's Aquila UAV was begun in 1979 and finally canceled in 1987 after its projected costs had more than doubled and it had met mission requirements on only seven of 105 test flights.)[13] As Table 1 indicated, the result has been numerous development efforts but few deployed systems. CBO has identified three main areas of concern in the Department of Defense's current unmanned aerial vehicle programs: the suitability of the tactical UAV (Outrider) for its intended missions, overlaps in capability, and issues related to the operational control of UAVs on the battlefield.

Suitability of Outrider for Its Missions

The cancellation of Hunter in January 1996 left the Pentagon without a new generation of unmanned aerial vehicles that could fly out to 200 kilometers and collect imagery. As noted above, the Joint Requirements Oversight Council decided to revise and enhance the specifications for the joint tactical UAV—which ultimately became the Outrider program—so that system could fulfill the various requirements of the Army and of the Navy and Marine Corps. But three major differences exist between those services' requirements.

First, the Army wants an inexpensive air vehicle that has a range of 50 to 60 kilometers and can support a brigade commander by collecting intelligence in his immediate vicinity. In contrast, the Navy and Marine Corps need a UAV system with a much longer range—200 kilometers—to support littoral operations.

---

13.   Statement of Louis J. Rodrigues, Director, Defense Acquisition Issues, National Security and International Affairs Division, General Accounting Office, before the Subcommittees on Military Research and Development and Military Procurement, House Committee on National Security, published as General Accounting Office, *Unmanned Aerial Vehicles: DoD's Acquisition Efforts*, GAO/T-NSIAD-97-138 (April 9, 1997), pp. 1-2.

Second, the Navy and Marine Corps want a vertical take-off and landing capability in their UAV system. The Navy would like its UAVs to operate from any ship capable of carrying a helicopter, but at the very least it wants the UAVs to be able to take off and land in 75 meters, or about one-third the length of a large amphibious ship. A UAV that can take off and land vertically would certainly provide that capability, but it would probably be much more expensive than the system the Army wants. Outrider will not have that capability.

Third, the Navy strongly prefers that a heavy-fuel engine—one that runs on diesel or jet fuel—power its new UAVs. The Army would also like that capability, but not as much as the Navy. Diesel fuel is far less volatile and combustible than gasoline, which is highly flammable. The Navy wants to remove gasoline from its ships altogether to reduce the risk of fire and explosion. Gasoline requires special preparations and handling, whereas diesel or jet fuel poses less risk. Nevertheless, Outrider, as currently planned, will run on gasoline.

Overlaps in Capability

A second major concern is an apparent overlap in the capability of what DoD calls the endurance UAV programs: Predator, Global Hawk, and Darkstar. According to the Department of Defense, each of those systems fills a niche in the pursuit of intelligence. But they also appear to create redundant capability that may not be necessary since the United States today faces a less threatening military environment than it did during the Cold War.

The medium-altitude Predator, which is already in production, can operate at 25,000 feet; as currently planned, Global Hawk will be able to operate at up to 65,000 feet, and Darkstar at up to 45,000 feet. Both Predator and Global Hawk can stay aloft for nearly 24 hours at their operating radius, but Darkstar is more limited at eight hours. Global Hawk can also carry a larger payload than Darkstar, which means a more capable sensor package. However, Darkstar's great advantage is that it is expected to be stealthy, so it can operate in areas with strong air defenses. The redundant capability derives from the fact that both Predator and Global Hawk may be survivable in threatening environments, though probably not as survivable as Darkstar.

Operational Control of UAVs on the Battlefield

Together, the Pentagon's UAV programs represent a relatively small amount of money: $620 million in 1999. Nevertheless, they sometimes find themselves at the center of debates about strategy and policy. Such is the case with Predator and the way in which DoD will fight the "deep battle"—that is, attacking enemy forces some 150 kilometers beyond the forward line of U.S. troops.

The Air Force believes it should be primarily responsible for engaging enemy forces in the area beyond the fire-support coordination line (the line separating ground forces' and air forces' responsibility for conducting a battle), which is usually about 30 kilometers beyond the forward line of troops. The Army, however, is pursuing the concept of "maneuver warfare" and increasingly views the deep battle as its responsibility. It would probably like to extend the fire-support coordination line to 150 kilometers. That debate over strategy has not been resolved and has indirectly entered into questions of how Predator will be controlled and assigned its missions.

Without Hunter, the Army no longer has a UAV that can support its efforts to engage enemy forces far beyond the forward line of troops. Consequently, it is looking to Predator to fill that gap. But the Predators now being acquired are owned and operated by the Air Force. The Army envisions equipping its corps and divisions with ground control stations and data terminals and having the Air Force lend them Predator air vehicles, which they would control directly. After the air vehicles had performed the tasks that the corps or division commanders required, or when they needed resupply, they would be handed back to their Air Force controllers.

The Air Force, however, argues that it must maintain operational control of the UAVs at all times to preserve the integrity of the airspace over the battlefield. Predators must be integrated into the "air tasking order" (the Air Force's guidelines for who does what) so that friendly forces do not end up shooting them down. The Air Force has stated that it will support Army requests to use Predator once higher priorities (as set by the theater commander) have been met. But the Army appears to doubt that the Air Force will be sufficiently responsive to its requests during battle.

To address some of the above problems, CBO has analyzed five possible alternatives for DoD's UAV programs. Those options, which could make unmanned aerial vehicles more effective, are presented in Chapter III. Before that, however, this paper examines the effects of the ACTD process on the development and cost of UAVs.

CHAPTER II

DEVELOPING UNMANNED AERIAL VEHICLES:

THE ACTD PROCESS

Programs to develop unmanned aerial vehicles for the U.S. military have frequently run into problems. In the mid-1990s, the Defense Airborne Reconnaissance Office, which is responsible for developing UAVs, tried a new approach: designing unmanned aerial vehicles as Advanced Concept Technology Demonstrations. The Predator medium-altitude endurance UAV was developed that way and is now being procured by the Air Force. Today, three other UAVs, the tactical Outrider and the high-altitude endurance Global Hawk and Darkstar, are being developed as ACTDs. Those programs prompt two questions: why has the Department of Defense had so much trouble developing UAVs? And has the use of the ACTD process improved that situation?

The second question is the more difficult of the two to answer. ACTDs are designed to develop new military systems and demonstrate them in the field faster and more cheaply than the traditional acquisition process. Developers in ACTD programs can use flexible management processes, "mature" technologies (ones that have already been proved in other systems), and close involvement by operational users and joint commanders to pursue those goals. Because of the complexities of UAVs, however, ACTD programs for those vehicles have had mixed success. Their rocky progress suggests that some of the causes of growing costs and delayed schedules are beyond the ability of the ACTD process to reform. But despite those problems, the ACTD approach appears to have had some success in areas where past UAV programs struggled, such as avoiding growth in operational requirements, improving cooperation among services and military commands, and providing commanders with the opportunity to try a new system in the field.

WHY DEVELOPING UAVs HAS BEEN DIFFICULT

Some Members of Congress and segments of the defense community have criticized DoD for its seeming inability to develop and field a tactical UAV. Recently, several Congressional committees nearly terminated Outrider in the budget authorization and appropriation process for fiscal year 1998. The problems that UAV programs have experienced stem from a number of different, but interrelated, factors.

First, many people, including defense contractors, appear to underestimate the difficulty of building an unmanned aerial vehicle system. For example, the Army Chief of Staff, General Dennis Reimer, was apparently mystified at the trouble

Outrider has experienced, saying "It's not laser brain surgery."[1] That may be true, but UAVs are still much more complex than the radio-controlled model airplanes they somewhat resemble. Building and operating a tactical UAV effectively involves integrating complex components into a small air frame. Those components include the engine, sensors, software, communications link, data link, avionics, ground control equipment, and so on. Making any one of those items work is not too difficult, but making all of them work together and still be much less expensive than a manned tactical aircraft is a major engineering challenge. One of the main problems afflicting the Outrider program is that the principal contractor drastically underestimated the technical challenges and level of effort required to produce a workable system that met both the cost requirement and most of the technical goals.

A second major problem among UAV programs is the tendency for "requirements creep." That is the phenomenon in which, once development has begun, the services impose more and more requirements on the UAV until meeting all of them becomes technically unworkable. In a broad overview of UAV programs, the General Accounting Office observed that requirements creep was a frequent occurrence. The Aquila UAV, for example, was supposed to have been a relatively simple, propeller-driven aircraft that could see over the next hill and relay imagery back to tactical commanders. But before its cancellation, the requirements had grown so much that it was expected "to fly by autopilot, carry sensors to locate and identify enemy point targets in day or night, use a laser to designate the targets for the Copperhead artillery projectile, provide conventional artillery adjustment, and survive against Soviet air defenses."[2] Outrider is suffering some of those same problems today; immediately before the start of the ACTD, its technical goals were expanded to cover the requirements that Hunter had been supposed to fill before it was canceled.

A third problem is that until recently, unmanned aerial vehicles have been the orphans of DoD, in that they have not had strong backing from the services compared with other priorities. That has been particularly apparent when problems have arisen in UAV programs, as they inevitably do in almost any development program. For example, near the end of its development, Hunter suffered three crashes in 45 days. The program was apparently terminated in part because the Army had a funding shortage in other areas that it considered higher priorities. Therefore, it helped cancel

---

1.   ". . . Unless Pilotless," Washington Outlook section, *Aviation Week and Space Technology*, November 10, 1997, p. 31.

2.   Statement of Louis J. Rodrigues, Director, Defense Acquisition Issues, National Security and International Affairs Division, General Accounting Office, before the Subcommittees on Military Research and Development and Military Procurement, House Committee on National Security, published as General Accounting Office, *Unmanned Aerial Vehicles: DOD's Acquisition Efforts*, GAO/T-NSIAD-97-138 (April 9, 1997), p. 2.

Hunter in the expectation that some of the funds would be redirected to fill those gaps (although that does not appear to have happened). In contrast, Pioneer—a Navy program—had the strong personal support of Secretary of the Navy John Lehman. He pushed and protected the program and ensured that it received adequate funding. Until Predator, Pioneer was the only UAV in the past decade to complete its development cycle and be fully deployed.

HOW ADVANCED CONCEPT TECHNOLOGY DEMONSTRATIONS WORK

DoD launched the Advanced Concept Technology Demonstration program in 1994 as an acquisition reform initiative.[3] The program has two main goals: to develop innovative military systems more quickly, and to involve the people who will actually use those systems on the battlefield in their development. The initiative is led by the Office of the Deputy Under Secretary of Defense for Advanced Technology (hereafter referred to as the Advanced Technology Office), which oversees several areas of emerging technology.

The ABCs of the ACTD Process

The ACTD concept occupies a unique niche among DoD's acquisition policies because of the close collaboration that it promotes between developers and operational users (or "warfighters") and because of its focus on mature technologies. Including operational users in the development process allows the military to assess the value of a new technology before acquiring it and develop new doctrine and tactics for that technology. By using the ACTD process, DoD hopes to reduce the time, and hopefully the cost, of producing a new weapon system and proving its military worth.

ACTDs also differ from traditional acquisition programs in schedule and scope. They are intended to be short (lasting no more than four years), small projects, costing anywhere from a few million dollars to a few hundred million dollars to complete. At their conclusion, ACTDs follow one of four paths: termination of the system, continued operation of only a few models, return of the system to the laboratory for continued development, or transition to procurement (see Figure 1).

The agency or service leading an ACTD is typically responsible for funding and executing the project, though the Advanced Technology Office occasionally

---

3.   See Congressional Budget Office, *The Department of Defense's Advanced Concept Technology Demonstrations*, CBO Memorandum (September 1998).

provides some funds. Much of the money for the UAV programs has come from the Office of the Secretary of Defense through the Defense Airborne Reconnaissance Office. That funding arrangement reflects the interest that high-level DoD officials take in those systems. It also means that reviews of the UAV programs include the Office of the Secretary of Defense as well as the service or agency participants.

An ACTD also has a user/sponsor, which is typically one of the joint unified commands. The U.S. Atlantic Command has been the user/sponsor for the high-altitude and Predator UAVs. The user is responsible for providing the military forces for the demonstration exercises in an ACTD. Those exercises are crucial to the outcome of an Advanced Concept Technology Demonstration, since a favorable review by a user can send a system to acquisition.

An explicit goal of the ACTD process is the development of joint systems. That process encourages the services to cooperate in managing projects and actively involves joint commanders in selecting ACTDs. Involving multiple services lets DoD use development funds efficiently in an era of tight defense budgets and also aids the development of joint warfighting strategies and tactics. Joint development

---

FIGURE 1. POSSIBLE OUTCOMES FOR AN ACTD PROJECT

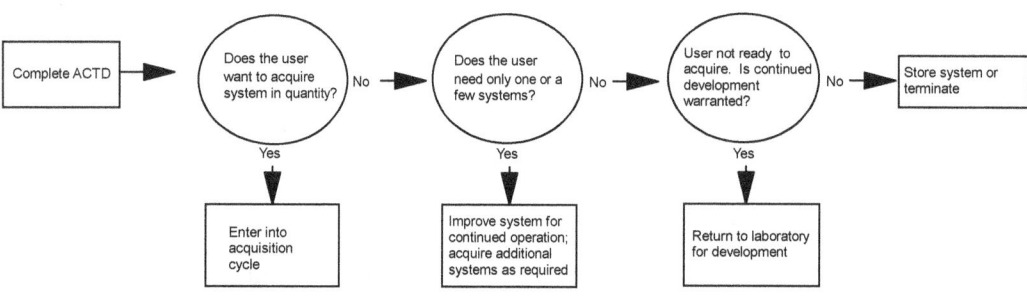

SOURCE: Congressional Budget Office based on information from Department of Defense, Defense Airborne Reconnaissance Office, *UAV Annual Report FY 1997* (November 6, 1997).

NOTE: ACTD = Advanced Concept Technology Demonstration.

of UAVs, however, precedes the advent of the ACTD approach. The Congress ordered DoD to establish the Unmanned Aerial Vehicles Joint Program Office in 1988 to manage the services' development efforts for all UAVs. (The Global Hawk/Darkstar program, which was started much later, is an exception; it is being run by the Defense Advanced Research Projects Agency, or DARPA.)

DoD and the Congress had hoped that a joint approach would end redundant UAV development programs by the different services and improve management. The jointly managed programs have struggled to succeed, however, with Predator being the only fielded system that was developed by the Joint Program Office.

Recent government actions suggest that, at least temporarily, the era of joint management of UAV programs may be waning. Frustrated again with problems in those programs, the Congress slashed funding for Outrider by over half in fiscal year 1998 and gave control of what remained to the Army. It also ordered DoD to transfer most of the Joint Program Office's responsibilities for managing UAV programs back to the services. In another move that may give more authority to the services, DoD has disbanded the Defense Airborne Reconnaissance Office, which had been serving as a central office for managing all of DoD's airborne reconnaissance systems. Many of the same functions will now be performed by the Office of the Deputy Under Secretary of Defense for Command, Control, Communications and Intelligence.

ACTDs as a Form of Prototyping

Although the ACTD process represents a new approach, it is a variation on an existing acquisition strategy, known as prototyping. The uses and kinds of prototypes vary widely, but a study by RAND defined the prototype acquisition strategy as building test models with the intention of learning more about a technology and, should the technology prove useful, reducing the future risks of development.[4] Essentially, an ACTD is a program that builds a technology demonstration prototype for users to evaluate. UAV prototypes are used for many of the same purposes as prototypes in other programs, such as to provide decisionmakers with better information for developing and acquiring systems or to be a hedge against uncertainty and risk in such development.

DoD has used prototypes to develop fighter aircraft, ships' combat systems, and missiles. How an acquisition program employs a prototype depends on the program's size and goals. DoD's interest in prototypes for the purpose of concept

---

4.   Jeff Drezner, *The Nature and Role of Prototyping in Weapon System Development*, R-4161-ACQ (Santa Monica, Calif.: RAND, 1992), p. 9.

demonstration has varied over the years, but their use in ACTDs represents a new application.[5] The dominant practice in the past has been to use highly integrated prototypes during the full-scale development of a system. In the ACTD program, the prototype is a means to a goal: the field demonstration.

Comparing examples of prototyping is difficult. Past studies have noted that the uses of prototypes vary so widely that virtually any comparison begs some caution. Besides variations in the prototyping strategy, changes imposed externally or internally to a system—such as changes in planned procurement quantity or performance goals—can affect the progress of development and the outcome of any system. Much of what ACTDs do has been tried in other programs. It is not so much that ACTDs have unique features, but that they are a blend of different methods for running a prototype program.

The unmanned aerial vehicle ACTDs share similar features with past prototyping programs, such as operational testing by service users and a flexible program-management philosophy. The prototype programs for the A-10 and F-16 aircraft, for example, both involved Air Force pilots early in the development phase of the aircraft.[6] The F-16 prototype program emphasized minimal documentation and left many of the program's details, such as performance objectives, up to the contractors. ACTDs also employ other acquisition reforms, such as making cost an independent variable and using integrated process teams to design products.

Three key features, however, make the ACTD approach different from past prototyping programs. First, the realistic nature of the experimental demonstration and the close cooperation between users and developers are unique to the ACTD process. The testing in an ACTD does not focus on technical testing to validate a system's performance, but instead emphasizes operational testing and field exercises to gather information on a system's performance from a user's point of view.[7] Second, the ACTD process allows the joint commander to be closely involved in developing a system, an opportunity that was not available in the past. Third, the degree of involvement by users is much greater than in previous prototype programs. The ACTDs, like other acquisition reform initiatives, seek to improve users' input into the development process and involve users in all aspects of the program.

---

5.   Ibid., p. 56.

6.   See Giles Smith and others, *The Use of Prototypes in Weapon System Development*, R-2345-AF (Santa Monica, Calif.: RAND, 1981).

7.   Because of the complexities of developing UAV systems, the UAV programs have involved the test and evaluation commands of the services and the Office of the Secretary of Defense.

Past studies have drawn no clear conclusions about whether prototyping helps meet cost, schedule, and performance goals. That should not be surprising, since the application of the strategy varies widely among acquisition programs, and a variety of external factors not directly related to prototyping frequently interfere in the management of a program.[8] Still, many of the ways that DoD plans to use prototypes in ACTDs have been useful in the past, and it hopes that they will help field new systems faster and more cheaply.

APPLYING THE ACTD APPROACH TO UNMANNED AERIAL VEHICLES

Are ACTDs appropriate for developing unmanned aerial vehicles? By employing that development approach, the Department of Defense is trying to steer the UAV acquisition strategy toward systems that support joint warfare, are focused on users' requirements, avoid development complications stemming from excessive requirements, and can be developed quickly. By avoiding lengthy acquisition programs and involving operators in demonstrations and development, the ACTD process is a way to educate users about the merits of a system and introduce realism into their expectations of it. Problems in the UAV programs, however, raise questions about the level of technical risk they really entail.[9]

How Risky Are the UAV ACTDs?

For ACTDs to meet their ambitious cost and schedule goals, the technologies they use must be mature enough to be developed quickly into an operating prototype. UAV program managers recognized that when they laid out their program goals. For example, the program manager for Predator wanted a first flight in six months, and the Outrider program set the same goal. Predator met its goal, but Outrider did not come close, a reflection of the programs' different level of maturity before the ACTD. Despite the ambitious goals of the ACTD approach, the myriad technical problems experienced by the current development programs seem all too similar to the problems of failed UAV programs in the past.

Until recently, the staff of the Advanced Technology Office did not adopt a set definition of technical maturity. They argued that case-by-case evaluation by a

---

8.   Drezner, *The Nature and Role of Prototyping in Weapon System Development*, p. 67.

9.   For a discussion of the technical problems with Darkstar, for example, see Department of Defense, Office of the Inspector General, *Audit Report on Advanced Concept Technology Demonstration*, Report No. 97-120 (April 7, 1997). For a discussion of technical issues in the development of Outrider, see General Accounting Office, *Unmanned Aerial Vehicles: Outrider Demonstrations Will Be Inadequate to Justify Further Production*, NSIAD-97-153 (September 1997).

review board was more effective than a strict definition.[10] Perhaps seeking to reduce some of the controversy that the ACTD process has attracted, the Advanced Technology Office finally provided a definition in the fall of 1997: any technology that is not in DoD's Basic or Applied Research Program or that has been demonstrated before the start of the utility assessment is considered mature. (The office grants exceptions if a technology that fails that test is not in a critical component of the system being developed.)

Although the requirement that ACTDs use only mature technologies is intended to minimize technical risk, critics could argue that this focus underestimates the risks of integrating new technologies into a working system. The highly capable UAVs desired by the services require using many leading technologies, such as advanced radar, composite materials, and stealth design features. In addition, the risks of integrating smaller subcomponents into a new system can be as great as developing a new technology. UAVs are a web of complex subsystems, and integrating the various payloads, software, air vehicles, and ground stations into a single system requires substantial expertise on the part of developers. The delays in the Outrider and Global Hawk/Darkstar programs can be traced in large part to such problems of integration.

In the context of ACTDs, the Advanced Technology Office views technical risk and integration risk as separate issues. That distinction is critical. Whereas the use of mature technologies is required for an ACTD, the acceptable level of integration risk is more flexible. According to the Advanced Technology Office, integration risks are addressed in the assessment of the overall risks of a program. But solving the integration problems of the Outrider and Global Hawk/Darkstar systems has proved more costly and taken longer than the Joint Program Office and DARPA had originally hoped.

Although most of the technologies that UAV programs use may meet the Advanced Technology Office's standard for a mature technology, each program's technical approach to developing a demonstration system is different. The two troubled programs, Global Hawk/Darkstar and Outrider, are new systems, with no predecessor to give their developers experience. In contrast, Predator was based on another system that was already in use, Gnat-750. The Predator program involved some amount of development, but Gnat-750 gave developers a working system on which to base their engineering efforts. Perhaps more important, by selecting a

---

10.   That review is done by the "Breakfast Club," a group of senior officials from the Advanced Technology Office, the Office of the Director for Defense Research and Engineering, and the Deputy Assistant Secretary of Defense for Command, Control, Communications, and Intelligence. The club also includes representatives from the Office of the Joint Chiefs of Staff, the Defense Advanced Research Projects Agency, the Ballistic Missile Defense Organization, and the science and technology and operational requirements offices of each military service.

proposal from the same contractor that developed Gnat-750 (General Atomics), DoD was working with a firm that had experience with UAV systems. Thus, although the smooth progress of Predator from development to deployment seems impressive, it probably reflects the maturity of the system that preceded it.

In developing the tactical and high-altitude endurance UAVs, DoD selected models that had no operational predecessors. Outrider was based on a paper system originally designed to meet a different requirement, and Global Hawk and Darkstar were entirely new designs. Difficulties with the design of the air vehicle and subcomponent integration have delayed the Outrider program by at least four months. Global Hawk has been delayed by about a year by software and development problems. Darkstar's flight-testing was postponed for over two years while designers improved the system. Those delays come as a disappointment to people who expected ACTDs to focus on demonstrating proven technologies.

It is not unusual for acquisition programs to encounter the sort of problems that Outrider and Global Hawk/Darkstar have run into. The experiences of those ACTDs suggests that, although the ACTD approach may have some practices that can improve the development process, many of the factors that cause costs to increase and schedules to slip may be beyond the scope of the reforms that DoD is pursuing as a part of the ACTD approach.

Is the ACTD Process Speeding the Delivery of UAV Systems?

The intent of all three UAV ACTDs was to put those systems in the hands of users as quickly as possible. Past UAV programs have languished in development, and few systems have ever made it to users. The amount of time spent in development is critical, because the Advanced Technology Office tries to limit ACTDs to only four years. Moreover, a protracted development schedule dilutes an ACTD's focus on operational evaluation. Because of differences in the maturity of the UAV systems, DoD has had mixed success in speeding deployment to the field. Predator moved relatively quickly to the field, but Outrider, Global Hawk, and Darkstar were soon overwhelmed by technical problems, delaying early demonstrations for their users.

Developing Criteria for Operational Assessment

In an ACTD, the lead service develops a draft concept of operation for a weapon system and the user/sponsor develops standards for evaluating that system in the field. The concept of operation is continuously refined as the ACTD progresses, reflecting changes as the system matures and as users refine their approach to operating it. Once a system is ready, users operate it in field exercises to assess its utility. That demonstration of military utility is at the heart of what makes an ACTD unique.

How much time is necessary for evaluating a system? That depends on how users plan to employ it. Predator participated in nine exercises over fifteen months to demonstrate aspects of the system's capabilities.[11] But in the Outrider and Global Hawk/Darkstar programs, the schedule for operational assessment has been shortened by the extra time required for development. Outrider completed a shortened 2½-month military-utility assessment. The high-altitude endurance program has reduced its scheduled utility assessment from about two years to one year.

Supporting the Joint Development of UAVs

The ACTD process has encouraged the joint development of UAV systems, but building such joint systems has proved more difficult than DoD expected. The Hunter and Medium Range UAVs were joint programs (although not ACTDs); both were hobbled by technical problems and management difficulties and were canceled before reaching full-rate procurement. For example, the payload that the Air Force developed for the Medium Range UAV would not fit into the air vehicle built by the Navy. That problem illustrated the importance of creating a means for effective project management among the many parties involved in developing a UAV. The ACTD framework provides just such a means for the current UAV programs and appears to be successful in coordinating the disparate demands of users, sponsors, and the services. But joint development can still complicate the task of creating a UAV system.

Developers can generally follow one of two approaches for joint systems. Either they can build a system that is operated by a single service but is interoperable with the data-dissemination systems of other services, or they can design a common system to be operated by two or more services. Experience with joint development suggests that the second option is the more challenging. Because the services have varying requirements for UAVs, both approaches have their advantages and disadvantages.

In an interoperable UAV system, one service takes charge of development, but other services make sure that their reconnaissance systems and computer networks can receive the UAV's data or imagery. Involving more than one service

---

11.   Michael Thirtle, Robert Johnson, and John Birkler, *The Predator ACTD: A Case Study for Transition Planning to the Formal Acquisition Process,* MR-899-OSD (Santa Monica, Calif.: RAND, 1997), p. 37.

in a program allows different users to determine whether the system meets their requirements. For example, both the Navy and the Army participated in developing Predator, which gave them an understanding of the system's capabilities and allowed them to evaluate it for their own needs. Developers made sure that Predator's sensors would perform well enough to meet operational needs and that its imagery could be distributed by joint dissemination systems. Similarly, the Global Hawk/ Darkstar program is designing its data-dissemination system to be interoperable with those of other services and joint commanders. The Air Force is the lead service on that program, but the Army and Navy are participating as well.

Interoperability has some drawbacks, however. When the demand for UAVs outstrips their availability—as has often been the case—the needs of tactical commanders may be sacrificed to those of higher echelons. That would probably not happen if the tactical commanders had their own, exclusive UAV systems. Another problem with interoperability is that one service can get stuck with the cost of providing UAV support to other services' commanders. If that leads the bill-paying service to monopolize the system's use or somehow diminish its availability, other commanders might have an incentive to demand their own systems or to forgo using UAVs altogether.

The other alternative, building a single UAV system for operation by several services, has the advantage of preventing redundant development efforts and giving each service's commanders their own system. However, to the extent that the services' requirements for a UAV differ, they must be willing to compromise to build a common system. If those requirements are too different, they may be unfulfilled or the system itself may be unworkable.

Problems with Outrider show that developing one system for many services can sometimes increase the technical challenges. That program is led by the Army, with the Navy and Marine Corps also participating. Outrider is in fact the only ACTD unmanned aerial vehicle that has pursued compatibility with naval operations as a system goal. Originally, the developers hoped to build one system for all of those services' tactical commanders to evaluate. However, technical delays have reduced the UAV's ability to meet the needs of some commanders and have postponed the delivery of demonstration systems to the field. For example, the technical goals for Outrider originally called for it to use a diesel engine to make it more suitable for shipboard operations. They also called for it to be able to land and take off from the Navy's large-deck amphibious ships. The diesel engine proved infeasible, however, and had to be dropped from the program. Those problems—and the decision not to test Outrider at sea—raise questions about how committed the Navy remains to the program.

THE EFFECTS OF THE ACTD APPROACH
ON THE COST AND CAPABILITY OF UAVs

Controlling the costs of acquiring and operating UAVs is a major concern for the agencies and services developing them. In the past, the services were unwilling to continue UAV programs when their costs grew beyond original estimates. (As a result, tactical commanders operating overseas now have only a few aging Pioneer systems to use, since the Aquila and Hunter programs were canceled.) In addition, some analysts have noted, the services were too ambitious in the capabilities that they demanded of past programs.

Balancing Affordability with Capability

Proponents believe that the ACTD approach makes development less expensive by reducing the time it takes. For example, the UAVs being developed as ACTDs forgo extensive flight testing and logistics analysis that normal acquisition programs carry out. During development, the lead service will conduct a logistics and maintenance analysis before assuming responsibility for the system, but that analysis is not supposed to be as significant an effort as would be carried out in a conventional acquisition program. That approach saves money during development but shifts the costs of a complete analysis to the post-ACTD period. If a system moves to acquisition, someone will have to go back and make those investments—as the Air Force is doing now with Predator.

Setting Flexible Goals for Capability

By not setting strict performance goals for UAV systems, and by allowing users to evaluate the systems early in the development cycle, ACTDs hope to avoid the requirements creep that has plagued other UAV programs. The short length of ACTDs and the policy of establishing early agreement about a system's concepts of operation limit opportunities for adding new requirements to a program. Program managers for Predator believe the ACTD approach helped keep the requirements for that UAV simple. The services have avoided adding other requirements to the Global Hawk, Darkstar, and Outrider programs during the ACTD, even though additional missions (beyond those the ACTDs will demonstrate) have been studied.

Moving Demonstrations to Acquisition

Because ACTDs are intended to be development efforts, which could have any one of a number of outcomes, they do not focus on many of the concerns that the acquisition process typically requires. For example, the acquisition process considers the production needs of the system and addresses its reliability and maintenance costs, whereas ACTDs generally do not. The UAV ACTDs are an exception, however, because the services are hoping to procure those systems. As a result, the developers need to worry about the transition process—moving the system that emerges from the ACTD to acquisition with minimal impact on the program's cost and schedule. The Global Hawk/Darkstar program is studying the costs of supporting and operating those systems as a part of the ACTD.

Problems with Predator's transition revealed weaknesses in the original ACTD's plans for preparing the system for acquisition. The Predator program office had achieved many of its demonstration goals, but several steps necessary for acquisition had not been required in the demonstration phase. For example, the Air Force had no technical manuals and insufficient data about Predator's reliability when it inherited responsibility for the system. Those had to be developed later, thus interrupting the program's progress.

Because of the difficulties with Predator's transition, the Advanced Technology Office has developed a more formal management approach to transition issues in ACTDs. The office now requires that an ACTD initially specify a lead service, which is responsible for operating and supporting the system after the demonstration is completed. In addition, managers in an ACTD now establish a special team to coordinate the transition. That increased emphasis on support and acquisition issues in an ACTD is not intended to imply a commitment to acquisition. Rather, by including concerns such as the life-cycle costs of a system in the ACTD, it is meant to keep developers from designing a system that costs too much to support once the demonstration is over.

A critical transition issue now confronting the Global Hawk/Darkstar program is how to support the manufacturing facilities. Cost overruns have forced DARPA to cut the number of air vehicles in that program, so a two-year gap now exists between the last planned air vehicle delivery (in 1999) and the first year any production deliveries could begin. According to the developer, Teledyne Ryan, that two-year gap will cause the loss of production workers and force the company to close the manufacturing facilities for Global Hawk. As a result, any production efforts that follow would have to pick up the additional costs of rebuilding the manufacturing base.[12] Similar transition issues are confronting other ACTDs, and the Advanced Technology Office is trying to provide them with funding.

Conclusions

Advanced Concept Technology Demonstrations have encouraged a different approach to developing unmanned aerial vehicles. Most of the advantages and disadvantages of that approach revolve around the merits of trying to develop new systems rapidly for demonstration. Demonstration is useful for learning about a system's combat effectiveness, but the progress of the UAVs shows that the ACTD process is prone to the same cost and schedule problems that can occur in the traditional acquisition cycle. Aspects of the ACTD process have prevented some of the pitfalls of past UAV programs. But the extensive development efforts those UAVs have required leave questions about the suitability of using that process to develop unmanned aerial vehicles.

12.　　　Michael A. Dornheim, "Global Hawk Begins Flight Test Program," *Aviation Week & Space Technology,* March 9, 1998, pp. 22-23.

CHAPTER III

ILLUSTRATIVE OPTIONS

FOR DoD'S UAV PROGRAMS

The Department of Defense's planned unmanned aerial vehicles, if fully developed and deployed, hold great potential for expanding the use of unmanned aircraft to collect intelligence. As noted in Chapter I, however, the programs to develop those UAVs involve various overlaps and problems.

The Congressional Budget Office has constructed five illustrative options to address some of the concerns that have been raised about those programs or to take greater advantage of the promise that UAVs appear to hold. All would use UAVs differently from the way DoD currently plans. However, the options by no means represent all of the possibilities for improving DoD's UAV programs. Nor are they motivated primarily by the desire to save money.

Option I focuses on DoD's highest priority for unmanned aerial vehicles: providing a UAV capability to the Army's brigade commanders. Within that option, the first alternative (Option IA) would provide that capability much more quickly than now planned by cancelling the Outrider tactical UAV and having brigade commanders use the existing Hunter systems instead. In addition, the Navy and Marine Corps would receive a new UAV more suitable for their maritime operations. The second alternative (Option IB) would keep Outrider but make it solely an Army system; the Navy and Marine Corps would continue to rely on their Pioneer tactical UAVs. Option II proposes a possible solution to the Army's concern that during a regional conflict its corps and division commanders would not receive support from Predator UAVs controlled by the Air Force. Specifically, Option II would give those commanders their own Hunters instead. Options III and IV examine the potential for trading off unmanned aerial vehicles for other aircraft: tilt-rotor UAVs for the Army's new Comanche reconnaissance helicopter (Option III), and Global Hawks for the Joint Surveillance Target Attack Radar System (Option IV), which the Quadrennial Defense Review has proposed reducing in number. Last, Option V addresses some Congressional concerns about overlaps in the UAV programs by cancelling Darkstar after it completes its Advanced Concept Technology Demonstration.

Many of the UAVs included in those options are still in the ACTD development phase, and DoD has not yet committed to buying them in quantity. As a result, the options cannot be compared with an overall DoD plan for unmanned aerial vehicles. Each option illustrates one possible alternative to one component of the department's UAV programs. Thus, it can only be compared with the plan for that particular option or the likely DoD plan once the UAVs finish development (assuming that the UAVs now under development—Outrider, Global Hawk, and Darkstar—are all ultimately successful and that the services buy and deploy them). Compared with those plans, two of the options would save money, two would cost money, and one would almost break even, based on both acquisition costs and operating and support costs over the assumed 15-year life of a UAV (see Table 5).

OPTION I: CANCEL OUTRIDER OR MAKE IT SOLELY AN ARMY SYSTEM

The Army has long had a need for a tactical UAV to support brigade commanders in combat operations. Indeed, the Joint Requirements Oversight Council considers the fielding of such a system its highest priority in the area of surveillance and reconnaissance. DoD would like to acquire 62 Outrider UAV systems, with the Army to get 38, the Navy nine, the Marine Corps 11, and four to be reserved for training. (The Navy systems are intended to have eight air vehicles each, whereas the Army and Marine systems would have four.) The Army is expected to procure one UAV system for each of its maneuver brigades—or three per division. A few divisions, the light ones, would receive four Outrider systems apiece. CBO estimates that acquiring those UAVs would cost about $860 million (in 1998 dollars), and operating and supporting them would cost an additional $930 million over 15 years.

Meeting DoD's goal has been difficult, however. Technical problems have delayed the Outrider development program, setting back the schedule for testing and producing the UAV. In addition, it is not clear that Outrider can satisfy the Navy's requirements for a tactical UAV, which include a vertical take-off and landing capability and a heavy-fuel engine to make shipboard operations easier. (The Navy is currently using Pioneer as a tactical UAV system that can be deployed on ships to support naval and littoral operations, but it had planned to phase out Pioneer in favor of Outrider.)

To address those problems with the Outrider program, CBO examined two alternative approaches. The first (Option IA) would cancel Outrider; instead, the Army would use Hunter for its brigade commanders, and the Navy and Marine Corps would procure a new tactical UAV. The second (Option IB) would keep Outrider but tailor it specifically to the Army. The Navy and Marine Corps would continue to use Pioneer.

Option IA: Use Hunter to Meet Army Brigade Requirements
and Buy Other UAVs for the Navy and Marine Corps

This alternative would quickly give the Army a UAV to support its brigade commanders by cancelling the Outrider program and taking full advantage of a

TABLE 5.   COSTS AND SAVINGS FOR FIVE ILLUSTRATIVE OPTIONS FOR UAVs
            (In millions of 1998 dollars)

	Acquisition Costs	15-Year Operating and Support Costs	Total
Option I: Provide a UAV Capability to Brigade and Task-Force Commanders			
DoD's Plan[a]	860	930	1,790
Cost of Option IA	780	1,020	1,800
Cost or Savings (-) Compared with DoD's Plan[b]	-80	90	10
Cost of Option IB	640	1,010	1,650
Cost or Savings (-) Compared with DoD's Plan	-220	80	-140
Option II: Provide a UAV Capability to Army Corps and Divisions Commanders			
Army's Plan[c]	0	0	0
Cost of Option II	250	500	750
Cost Compared with Army's Plan	250	500	750
Option III: Trade Off UAVs for Reconnaissance Helicopters			
Army's Plan[d]	31,500	6,600	38,200
Cost of Option III	27,700	6,000	33,700
Savings Compared with Army's Plan	-3,800	-700	-4,500
Option IV: Supplement JSTARS Coverage with UAVs			
Air Force's Plan	1,700	4,300	6,000
Cost of Option IV	2,200	5,000	7,200
Cost Compared with Air Force's Plan	500	700	1,200
Option V: End Darkstar Production with the ACTD Vehicles			
Air Force's Plan[a]	2,600	1,900	4,600
Cost of Option V	2,000	1,600	3,600
Savings Compared with Air Force's Plan	-600	-400	-1,000

SOURCE: Congressional Budget Office.

NOTE: UAV = unmanned aerial vehicle; DoD = Department of Defense; JSTARS = Joint Surveillance Target Attack Radar System; ACTD = Advanced Concept Technology Demonstration.

a.   CBO's assumed plan based on available information.

b.   The new UAV for the Navy and Marine Corps represents about 60 percent of these costs. If one were to compare the Army component only and assume Outrider is procured as an Army-only system, buying Hunter instead of Outrider would save about $400 million in total costs.

c.   The Army plans to use Predators bought and operated by the Air Force, so they will cost the Army nothing.

d.   The costs of the Army's plan for Option III are based on the full Comanche program of 1,292 helicopters, not just the number used in cavalry troops.

system that already exists: Hunter. Many Hunter air vehicles and their support equipment were purchased by the Army during the mid-1990s and are still in storage. At the same time, the Navy and Marine Corps would buy a new UAV with vertical take-off and landing capability to better meet their unique requirements. (The Navy would acquire nine of the new UAV systems and the Marines 11—the same numbers as planned for Outrider. Two additional systems would be purchased for training.) Compared with buying and operating 62 Outrider systems, Option IA would save $80 million in acquisition costs but add about $90 million to total operating and support costs (see Table 5).

Meeting the Army's Brigade Requirements with Hunter. The Army owns 56 Hunter air vehicles and 28 ground control stations as well as other support equipment. They were purchased in seven systems, with eight air vehicles each, as part of the low-rate production contract in 1994 and 1995. Currently, one Hunter system is stationed at Ft. Hood in Texas to support Task Force XXI—the Army's new digitized warfighting brigade. A second system (with four air vehicles) is stationed at Fort Huachuca in Arizona for training purposes. The other systems are in storage.

Equipping the Army's maneuver brigades with Hunter would require having an additional 18 air vehicles and 33 mobile UAV support team (MUST) packages, which consist of a modern ground control station and a smaller ground data terminal for immediate deployment. This alternative would also buy 82 air vehicles to allow for peacetime attrition (see Box 1). Those extra purchases—100 air vehicles and 33 MUST packages—are in addition to the Hunter equipment already in the Army's inventory. The current ground support equipment would be deployed with the division headquarters, and the new MUST packages would be assigned to the brigades.

This alternative would take advantage of what has been called the "pass-forward" method of operating and supporting unmanned aerial vehicles. As would be the case with Outrider, a Hunter system would be attached to a division's military intelligence battalion and would comprise six air vehicles, the principal ground control stations, and most other support equipment. Divisions are better able to support the logistics required by a Hunter system than individual brigades are. However, each brigade would have a MUST package with which it could take direct control of a Hunter air vehicle. The air vehicle would launch from the rear near the division headquarters and would be "passed forward" to the modern ground control station with the brigade. The brigade commander could then use the air vehicle for up to 10 hours before it needed to land for refueling and other support. He would have direct control of the air vehicle until he passed it back to the military intelligence battalion of the division, where its primary support base would be located.

BOX 1.
CALCULATING ATTRITION FOR UAVs

When the U.S. military buys weapon systems or other equipment, it frequently purchases an extra amount to make up for expected attrition—damage or losses that occur in testing or routine use—during peacetime. (Wartime attrition is unpredictable and so is not generally planned for in acquiring a weapon system.) For example, when the Army buys a helicopter, it purchases an additional 10 percent to replace the small portion of its inventory (less than 1 percent) that it loses to attrition every year. In the case of unmanned aerial vehicle (UAV) programs, determining how many air vehicles will need to be purchased to make up for attrition over the life of a program is a difficult and imprecise exercise. Only one UAV, Pioneer, has made it through development, acquisition, and more than 10 years in the field. Thus, military planners have only limited experience on which to base their estimates of attrition. Furthermore, unmanned vehicles appear, so far, to be much less reliable than manned aircraft. Overall, the bugs have generally been worked out of manned systems.

In analyzing the options in this chapter, the Congressional Budget Office (CBO) based its estimates of attrition for tactical UAVs on the expected attrition purchases of Outrider.[1] (CBO used that common standard for the sake of simplicity and to avoid biasing arguments for or against a particular alternative.) Current plans for Outrider call for buying 209 attrition air vehicles between fiscal years 1998 and 2008 to supplement the 284 vehicles that would be deployed with troops or used for training. That expected attrition represents about three-quarters of the entire Outrider program over a 10-year period, or about 7 percent a year.

Actual attrition may be significantly greater or less than that amount. Pioneer, for example, suffered about 75 peacetime crashes out of nine systems (with five air vehicles each) over 10 years—representing an attrition rate more than double that planned for Outrider. However, expecting a lower peacetime attrition rate than Pioneer's for future UAVs may not be unreasonable. For example, most tactical UAVs will eventually have Common Automatic Recovery System (CARS) hardware and software incorporated into their avionics. If it works as expected, CARS will guide the air vehicles in landing—one of the more risky maneuvers they perform—thus reducing the potential for human error. Furthermore, Pioneer was not acquired with maintainability or reliability in mind. The Navy's objective in the 1980s in developing Pioneer was to get a UAV system to its fleet quickly. In the process, it sometimes bypassed normal acquisition rules and procedures.

---

1.   The one exception is Bell Helicopter's tilt-rotor UAV, which is discussed in Option III. CBO assumed that a smaller number of air vehicles could be bought for attrition because every system on that air vehicle has a backup (at considerable extra expense). However, in that option, the tilt-rotor is not being compared with other tactical UAVs but with the Comanche helicopter, which is expected to have a much smaller attrition rate.

Time to Full Operational Capability. Substituting Hunter for Outrider could shave several years off the time needed to give brigades their own UAV capability (see Table 6). According to the current schedule, Outrider should be deployed to all of the Army's maneuver brigades by 2004. However, it has experienced a number of development problems that the contractor may or may not be able to resolve. History suggests that developing UAVs involves a learning curve, and there may be more bumps in the road before Outrider is a mature system.

Hunter, by contrast, is already a mature, reliable system. The Army has used it successfully in various training exercises and in developing concepts of operation for tactical UAVs. The existing Hunter systems could be brought out of storage and deployed in much less time than waiting for Outrider. Those systems would need some upgrades and improvements, but that would not take long. The real bottleneck would be an insufficient number of trained crews. Currently, the Army can train two crews a year. It could expand that, but at some additional expense.

TABLE 6.   MEETING ARMY BRIGADE-LEVEL UAV REQUIREMENTS UNDER
             DoD'S PLAN AND OPTION IA

	DoD's Plan (Outrider)	Option IA (Hunter)
Time to Full Operational Capability (Years)	5	Less than 2
Capabilities		
Dash speed (Kilometers per hour)	204	196
Number of sorties required for 24-hour operations	8	3
Coverage (Square kilometers per hour)	61[a]	106[b]
Payload (Pounds)	65	200
Deployability (Number of C-130 sorties required)	4	8

SOURCE: Congressional Budget Office based on data from the Department of Defense.

NOTE: UAV = unmanned aerial vehicle; DoD = Department of Defense.

a.   Assuming that the air vehicle is flying at an altitude of 5,000 feet and a speed of 140 kilometers per hour and using only the electro-optical sensor.

b.   Assuming that the air vehicle is flying at an altitude of 5,000 feet and a speed of 165 kilometers per hour and using only the electro-optical sensor.

Capabilities. Hunter provides equal or greater capability to brigade commanders than Outrider would in four areas: dash speed, endurance, coverage, and payload. Speed is arguably the most important capability that a brigade commander wants in an air vehicle. Dash speed (the maximum speed at which a vehicle can travel for short periods) represents the responsiveness of the UAV assigned to a particular unit and determines how quickly it can provide intelligence at the right time and in the right place. Outrider and Hunter have essentially the same dash speed—around 200 kilometers per hour.

Endurance is another measure of capability in which this option exceeds the Army's plan. Based on the number of hours that the UAVs can operate at 200 kilometers (Outrider's intended radius), Hunter's endurance is more than twice that projected for Outrider—eight hours versus three hours. Thus, a brigade commander could study, track, or cover a particular target or area for a much longer time with Hunter than with Outrider. Another way of looking at that measure is the number of sorties required to provide continuous 24-hour imagery collection for each brigade in a division. The Army's likely plan for using Outrider would require eight sorties per brigade to provide nearly 24-hour operations. Option IA would require three Hunter sorties for the same capability.

Coverage is the amount of area a UAV system can cover with its sensor. That area depends on the air vehicle's altitude, camera, and endurance and on the number of vehicles that can be simultaneously controlled in the air by the available ground stations. This option provides a division's brigades with greater coverage than the expected Army plan—106 square kilometers per hour versus 61—largely because of the Hunter system's greater endurance and larger number of vehicles that can be controlled in the air.

Finally, Hunter's potential payload, 200 pounds, is more than three times that expected for Outrider. Hunter carries a similar sensor but also has room to carry additional payloads that are now being tested or developed, such as a laser to guide precision munitions to their targets. Outrider may be able to carry other payloads as well, but with a smaller air frame, it is not likely to do so as easily as Hunter.

Deployability. Critics of using Hunter to support brigade requirements have pointed to the system's large "logistics footprint" and deployment requirements. A full Hunter system of eight air vehicles and four ground stations originally required five-ton trucks and numerous sorties by C-130 aircraft to deploy it. By contrast, an Outrider system is supposed to be deployable in a single C-130 sortie (although in reality it will require slightly more than that).

The Hunter system that this option envisions for a division's brigades—six air vehicles, three ground control stations, accompanying shelters, data terminals, and three MUST packages—would take about eight C-130 sorties to transport to a theater of operations. Deploying a division's three Outrider systems (one for each of the maneuver brigades) is expected to require at least four C-130 sorties. Thus, the deployment requirements to support the three brigades of a division would be twice as great under this option as with the planned Outrider systems.

Overall Assessment. The option of using the existing Hunters in the Army's inventory, as well as purchasing a few more, would yield significantly greater UAV capability for brigade commanders faster and at a lower cost than the Army's likely plan. The chief disadvantage of this option is its greater deployability requirements. In addition, replacing air vehicles lost to attrition would be much more expensive because Hunter is a more capable platform. Overall, including those "attrition spares," Hunter would cost less to acquire but more to support than Outrider.

Meeting the Navy's and Marine Corps's UAV Requirements. Option IA would meet Navy and Marine requirements by purchasing a UAV with vertical take-off and landing capability and a heavy-fuel engine. The Navy has been examining several such systems for possible purchase. A typical example is Guardian, built by Cana-dier. The company has been flying a smaller-scale demonstrator model of Guardian for at least 10 years. The full-production model employs many of the same com-ponents as the demonstrator but has a larger air frame and rotors.

With a greater payload and range, such a UAV appears to be much better suited to meet the Navy's and Marine Corps's needs than Outrider, principally because of its vertical take-off and landing capability and its heavy-fuel engine. The principal disadvantages of such systems are that they have not yet had much flight-testing, and they are likely to be more expensive than Outrider.

Time to Full Operational Capability. In terms of development time, there is little to distinguish between this option and DoD's plan (see Table 7). The time needed to finish modifying a demonstrator model to bring it up to production specifications, integrate it with the tactical control system, and begin production is around two and a half years, at a minimum. Outrider is supposed to complete its development process and begin low-rate production in fiscal year 1999, which would give it the edge in development time, but its past problems lend little confidence to predicted schedules. Of course, a new UAV for the Navy and Marine Corps could experience delays as well. In addition, the services would then require several more years to buy and deploy the systems in the field.

Capabilities. With the exception of its ability to take off and land vertically, the capabilities of a typical maritime UAV are fairly similar to those of Outrider. CBO assumed that each naval task force or amphibious ready group would receive one UAV system—composed of a ground control station, four air vehicles for a Marine system or eight air vehicles for a Navy system, and associated data terminals and ground support equipment. Outrider's dash speed is expected to be faster than that of a typical UAV with vertical take-off and landing: around 200 kilometers per hour versus 160. (Some systems under development that have vertical take-off and landing could fly much faster than that, but they would probably be more expensive than the option CBO considered.) However, Outrider's endurance is expected to be slightly less; thus, eight Outrider sorties would be required for 24-hour operations, compared with about seven sorties for a UAV with vertical take-off and landing. CBO could not compare the coverage of the two systems because data for the prospective maritime UAV system were not available.

A significant area of difference between the two systems is likely to be their payloads. Whereas Outrider is expected to carry a 65-pound payload, a typical maritime UAV with vertical take-off and landing can carry 110 pounds. That gives it a somewhat greater potential to handle the type and quantity of payloads the nautical services may want their UAVs to carry.

TABLE 7.   MEETING NAVY AND MARINE CORPS UAV REQUIREMENTS UNDER DoD'S PLAN AND OPTION IA

	DoD's Plan (Outrider)	Option IA (Maritime UAV)
Time to Full Operational Capability (Years)	5	5
Capabilities		
Dash speed (Kilometers per hour)	204	157
Number of sorties required for 24-hour operations	8	7
Coverage (Square kilometers per hour)	61[a]	n.a.
Payload (Pounds)	65	110
Deployability (Number of C-130 sorties required)[b]	1 to 2	2

SOURCE: Congressional Budget Office based on data from the Department of Defense.

NOTES: Navy UAV systems would have eight air vehicles and Marine systems would have four.

UAV = unmanned aerial vehicle; DoD = Department of Defense; n.a. = not available.

a.   Assuming that the air vehicle is flying at an altitude of 5,000 feet and a speed of 140 kilometers per hour and using only the electro-optical sensor.

b.   This applies only to the Marine Corps, since Navy UAVs would already be deployed aboard ship.

Deployability. The deployability of either system is not an issue if the UAVs are stationed on board ships before a task force or amphibious ready group leaves port. Marine systems might need to be airlifted to an overseas location, however. Both an Outrider and a naval UAV system are expected to fit on a C-130 and would require at most two sorties for delivery to a theater.

Overall Assessment. The main concern about Outrider is its suitability for ship and fleet operations. The chief advantage of this option is that it provides an unmanned aerial vehicle system that is highly suitable to shipboard operations, primarily because of the UAV's vertical take-off and landing capability and heavy-fuel engine. Otherwise, the capabilities of the two systems are fairly similar. Outrider will almost certainly be the cheaper system, but money saved by using Hunter to fulfill the Army's brigade UAV requirements could offset the additional costs of a naval UAV.

Option IB: Buy Outrider Only for the Army

This alternative addresses the same issues as Option IA but in a way that favors the Army's UAV needs over those of the Navy and Marine Corps. It proposes buying the Outrider tactical UAV for Army use, while the Navy and the Marine Corps would continue relying on their Pioneer systems. The upgrades currently planned for Pioneer would continue, and replacements for air vehicles lost through attrition would be purchased as needed to maintain the existing systems.

Some of the problems that have confronted the Outrider program stem from the Navy's requirements for the UAV, which include a 200-kilometer range, a heavy-fuel engine, and the integration of extra components for shipboard operations.[1] Eliminating those requirements would leave a system capable of fulfilling most Army requirements. In fact, one high-ranking Army official has stated that Outrider—even with its problems—appears able to meet the Army's needs and should therefore be procured.

In terms of the Army, this option would be the same as the service's likely plan for Outrider. Thus, all of the relevant operational factors—time to full operational capability, deployability, and capabilities—would be the same as under that plan. For the Navy and Marine Corps, this option would represent no change from their current UAV situation. Pioneer is roughly as capable a system as Outrider, except that it is over 10 years old, requires a great deal of maintenance, and is more difficult to work with. Of course, there is no guarantee that Outrider would be able to overcome its development problems and emerge as a useful system for shipboard

---

1.   See, for example, General Accounting Office, *Unmanned Aerial Vehicles: Outrider Demonstrations Will Be Inadequate to Justify Further Production*, NSIAD-97-153 (September 1997), pp. 4-7.

operations. This option would spend less on acquisition than either Option IA or DoD's plan. However, those savings would be partially offset by higher operating and support costs over 15 years (compared with DoD's plan). As a result, net savings from Option IB would be around $140 million compared with DoD's plan and $150 million compared with Option IA. In spite of those overall savings, the Army's Outrider systems would have a higher cost per system (unit cost) because this option would buy fewer of them.

OPTION II: USE HUNTER TO MEET THE ARMY'S DIVISION AND CORPS UAV REQUIREMENTS

Option II attempts to avoid the problems that could arise if the Army relied on Predator unmanned aerial vehicles controlled by the Air Force to meet its division and corps UAV requirements. With Hunter terminated, the Army proposes to rely on Predator—an Air Force system—to handle UAV missions at the division and corps level. However, the Air Force plans to buy just 12 Predator systems (each with four air vehicles and one ground control station) and to deploy only five of them to a regional conflict. The Air Force has stated that although it is willing to use Predator to support division and corps commanders, there may be higher priorities set by the theater commander or the national command authority that could require most, if not all, of the Predator assets. If two corps and seven divisions deployed to a regional conflict—as happened during the Gulf War—it seems unlikely that the average division commander would get a prompt response to his request for a Predator to perform a reconnaissance mission.

One possible solution to that problem is to provide each division and corps with its own UAV capability using the Hunter systems that the Army has in storage. This option would give a Hunter system with six air vehicles and three ground control stations to each corps and a Hunter system with four air vehicles and two ground control stations to each division—for a total of 64 air vehicles and 32 ground control stations. The Army already owns 56 air vehicles and 28 ground stations, so this option would require only a small purchase to fill out the force and to provide some extra systems for training and for replacing those lost through attrition.

Those Hunters are the same ones that would be given to the Army's brigade commanders under Option IA. As a result, Options IA and II could not be pursued simultaneously without buying substantially more Hunter systems.

Costs

Relying on the Air Force's Predators to provide imagery to corps and division commanders would cost the Army nothing. The purchase of those UAVs is already planned and included in the Air Force's budget, so giving them an extra mission to perform during a regional conflict essentially entails no added costs.

This option, by contrast, would increase the Army's costs. It would require a total of 72 air vehicles and 36 ground control stations. Bringing the existing Hunter systems out of storage would cost little, but buying 16 additional air vehicles and eight ground stations (plus attrition spares)—as this option envisions—would mean an additional $250 million in acquisition costs and about $500 million in operating and support costs over 15 years for all systems.

If the alternative were for the Army to procure its own Predators—something it has no plans to do—this option would save money in comparison. Buying a Predator system for every corps and division (14 additional systems in all) and operating them for 15 years would cost about $1.7 billion.

Capabilities

Both the Army plan and Option II have various strengths and weaknesses in the area of capability. The most pronounced strength of the Army's plan to use Predator is that system's 24-hour endurance. Only one Predator sortie is necessary to provide 24-hour coverage of a particular area or target. However, should a division commander ever need such lengthy coverage, he would probably be unlikely to get one of the few available Predators assigned to him for that long. By comparison, a Hunter system attached to a division would need to use three sorties for 24-hour operations (see Table 8).

The most pronounced strength of Option II is Hunter's dash speed. The Hunter systems assigned to a corps or a division could fly at almost 200 kilometers per hour, compared with 130 kilometers per hour for Predator. Furthermore, corps and division commanders would have more than one Hunter air vehicle at their disposal; thus, in a sense, the flexibility of their system would be multiplied by the number of vehicles they could put into the air simultaneously. Even if such commanders could automatically receive a Predator from the Air Force, they would still have much greater flexibility with their own Hunter system. Overall, those capabilities make Hunter more responsive to the immediate needs of its users.

In terms of payload, although a Predator air vehicle has much greater capacity than a Hunter, the total amount of payload available to unit commanders under the two alternatives would be more comparable. Under the Army plan, a corps or division commander would control a single Predator air vehicle and thus could use payloads of up to 450 pounds. But under this option, a corps commander would potentially have three air vehicles, each carrying 200 pounds, at his disposal. A division commander would have two air vehicles at 200 pounds each. Thus, although corps or division commanders could take advantage of a heavier payload under the Army's plan, under this option they would have the flexibility to put different payloads on different air vehicles. For example, because a corps commander could simultaneously control three air vehicles, one could contain a conventional imagery payload, another could have a laser designator, and a third could carry a countermine payload.

TABLE 8.   MEETING CORPS- AND DIVISION-LEVEL UAV REQUIREMENTS
            UNDER THE ARMY'S PLAN AND OPTION II

	Army's Plan (Predator)	Option II (Hunter)
Time to Full Operational Capability (Years)	5	Less than 2
Capabilities[a]		
Dash speed (Kilometers per hour)	130	196
Number of sorties required for 24-hour operations	1	3
Coverage (Square kilometers per hour)	104[b]	110[c]
Payload (Pounds)	450	200
Deployability (Number of C-130 sorties required)	5	8 for a corps, 5 for a division

SOURCE: Congressional Budget Office based on data from the Department of Defense.

NOTE: UAV = unmanned aerial vehicle.

a.   The worst-case scenario—no responsiveness from the Air Force's Predator systems—would mean zeros for this category.

b.   Assuming that the air vehicle is flying at an altitude of 10,000 feet and a speed of 130 kilometers per hour and using only the electro-optical sensor.

c.   Assuming that the air vehicle is flying at an altitude of 10,000 feet and a speed of 165 kilometers per hour and using only the electro-optical sensor.

Deployability

The UAV systems under the Army's plan and this option are roughly comparable in terms of deployability. One Predator system requires five C-130 sorties to deploy. A Hunter system of six air vehicles requires about eight C-130 sorties, and one with four air vehicles needs five (see Table 8). Predator deployments are often described using a different measure, however: the number of C-141 aircraft. A Predator system requires two such sorties for deployment, whereas a Hunter system with six air vehicles would require about 3½ C-141 sorties, and a system with four would need less than 2½ sorties.

Delivering five Predator systems to support a major regional conflict, as the Air Force plans, would require 10 C-141 sorties. If the Army deployed two corps and seven divisions to the theater, as it did during the Gulf War, then the Hunter systems accompanying those units would need another 23 C-141 sorties. Those sorties would have to be counted as additional to the ones required for Predator because the Predator systems would almost certainly be deployed to the theater even if the Army had its own UAV capability for its corps and divisions.

Overall Assessment

Compared with the Army's plan, the most significant benefit of this option to corps or division commanders would be having their own unmanned aerial vehicle capability. They would not have to wait in line for a Predator to become available. Instead, with a Hunter system already available, they would be able to see what they wanted to see as soon as the air vehicle could reach the target area.

The chief disadvantages of this option relative to the Army's plan are the cost and the substantial additional requirements for deployment. Regardless of whether the Army began using Hunter, the Air Force would still want to procure Predator. Thus, the Hunter systems would be added to the equipment that would need to be delivered to a major regional conflict, requiring an extra 23 flights by C-141 aircraft. This option would also be more expensive for the Army. Even though many of the Hunter air vehicles and ground stations are already in the Army's inventory, pulling them out, making them ready for combat, and buying several more would entail costs to the service that the Air Force's procurement of Predator would not.

OPTION III: BUY TILT-ROTOR UAVs AND REDUCE THE ARMY'S PLANNED COMANCHE HELICOPTER FORCE

Many people would agree that unmanned aerial vehicles promise to enhance the fighting potential of U.S. forces on the battlefield by giving commanders immediate information about the disposition of enemy troops. After the Army's warfighting experiments in 1996 at the National Training Center, the Army Vice Chief of Staff described tactical UAVs as "major combat multiplier[s]" for a brigade commander.[2] But compared with the combat power of an aircraft, for example, it is difficult to quantify the role that reconnaissance plays in warfare. Doubling the number of missiles that an aircraft can carry will double its combat power and, thus, the number of potential targets it can attack. But introducing new or better reconnaissance systems, although almost certainly making military forces better off, is not such a straightforward combat multiplier.

This option illustrates one way in which the military could make even more use of UAVs than it now plans. The option assumes that having many UAVs operating on a battlefield is indeed a major combat multiplier—in other words, that their reconnaissance capability will enhance the fighting power of U.S. forces. Consequently, this option would substitute Eagle-Eye—a tilt-rotor UAV developed by Bell Helicopter that the Navy has been examining—for some of the Comanche reconnaissance helicopters that the Army plans to buy. The Army's aviation forces would receive 369 tilt-rotor UAV systems (each with one ground control station and three air vehicles) in lieu of an equal number of Comanches. Taking into account attrition spares and "maintenance float" (extra aircraft bought to be used while others are undergoing maintenance), that would mean reducing the planned Comanche acquisition by 461 aircraft and purchasing a total of more than 1,900 Eagle-Eye air vehicles (see Table 9).

Several years ago, the Department of Defense ordered the Army to look into how Comanches and UAVs could be used together. The Army has a study under way to examine and test the concept of having a Comanche helicopter control a tactical UAV on the battlefield. But the service was also specifically asked to examine potential "trade-offs" between reconnaissance helicopters and unmanned aerial vehicles. It is not clear whether the Army is seriously examining that question. This option provides a brief look at the advantages, disadvantages, and costs of replacing reconnaissance helicopters with UAVs.

---

2.   Gen. Ronald H. Griffith, "Memorandum for Under Secretary of Defense for Acquisition and Technology on Tactical Aerial Vehicles" (May 8, 1997).

TABLE 9.   NUMBER OF UAVs SUBSTITUTED FOR COMANCHES UNDER OPTION III

	Comanches Not Bought	UAVs Bought Instead
Number Deployed with Forces		
Helicopters or Air Vehicles	369	1,107
Ground Control Stations	*	369
Number Used for Training		
Helicopters or Air Vehicles	0	108
Ground Control Stations	*	36
Number Used for Attrition Spares and Maintenance Float[a]		
Helicopters or Air Vehicles	92	715
Ground Control Stations	*	81
Total		
Helicopters or Air Vehicles	461	1,930
Ground Control Stations	*	486

SOURCE: Congressional Budget Office.

NOTE: UAV = unmanned aerial vehicle; * = not applicable.

a.   Extra aircraft to replace those lost, damaged, or undergoing maintenance.

How Would It Work?

Substituting unmanned aerial vehicles for Comanches in future Army units could be done in one of two ways—both of which involve cavalry squadrons, regimental aviation squadrons, and corps target-acquisition and reconnaissance companies and platoons. The Army intends for the typical cavalry squadron attached to a division to include two troops of 12 RAH-66 Comanche helicopters.[3] One way to carry out Option III would be to substitute one troop of 12 UAV systems for one troop of 12 Comanches. The UAVs would be controlled from the ground, and the helicopters and UAVs could operate separately or together. If they operated together, the

---

3.   Under the Army's reorganization of its aviation assets—the Aviation Restructure Initiative—the two air cavalry troops per squadron will have either eight OH-58D Kiowa Warrior helicopters or eight AH-1 Cobra helicopters. When the Comanche is fully fielded, each troop is intended to have 12 of them. Thus, compared with the Army's current force structure, this option would reduce the number of armed helicopters only slightly and provide a large additional reconnaissance capability.

helicopters could receive reconnaissance information from the UAVs via a communications link from the ground control stations, or the helicopters could themselves carry imagery data terminals that would allow them to see what the air vehicles see.

A second, more experimental way to implement this option would be to substitute six unmanned aerial vehicle systems for six helicopters in each troop. Then, each helicopter in a troop could control one UAV. The Army study that is looking at the technical challenges of controlling an unmanned aerial vehicle with a helicopter has not reached a conclusion about its feasibility. Having a helicopter pilot send out a UAV on a mission and receive nearly instantaneous imagery would avoid the need to operate through the communications link with the ground control station, which could save precious time in a tense combat situation. However, it also risks overloading the already busy helicopter pilot.

According to the Army doctrine, cavalry forces have two primary missions: reconnaissance and security. Reconnaissance includes route, area, and zone reconnaissance missions, and security includes screening, guarding, and covering other forces. Army officials say UAVs are capable of performing the reconnaissance missions. But in most cases, helicopters are better suited for supporting security missions, because such missions typically involve a much greater likelihood of combat.

Cavalry squadrons would not be the only forces affected by Option III. Target-acquisition and reconnaissance companies (or platoons) are assets of divisions and corps and are used mainly for supporting those units' artillery. They help identify targets, correct fire, and assess damage. In this option, tilt-rotor UAVs would also be substituted for all of the helicopters (a total of 93) assigned to those missions.

Costs

A principal advantage of Option III is cost. Tactical UAV systems, even the most sophisticated, are less expensive than manned aerial platforms, including Comanche. Substituting tilt-rotor UAV systems for Comanches would save the Army about $3.8 billion in acquisition costs. It would also save about $700 million over 15 years of operation and support, for a total savings over the life of the UAVs of about $4.5 billion.

Loss of Firepower

The biggest drawback of Option III is that unmanned aerial vehicles are not armed reconnaissance helicopters. To some degree, this option represents an apples-to-oranges comparison. The Eagle-Eye tilt-rotor UAV does not carry weapons, partly because it is unmanned and relatively cheap so losing it is not particularly worrisome.[4] The more expensive Comanche, by contrast, has weapons for self-defense, as well as attack weapons such as Hellfire missiles. Thus, units that lost helicopters and gained UAV systems under this option would lose a substantial percentage of their firepower. For that reason alone, the Congress may not find the option feasible.

The trade-off CBO is emphasizing in this option is between platforms to conduct reconnaissance, surveillance, and target acquisition. The chief disadvantage of the UAV in that arena is that, unlike Comanche, it does not carry a pair of human eyes on board, which in some circumstances can be irreplaceable to confirm the identification of a particular target or reconnoiter a particular spot. Overall, however, the tilt-rotor UAV has many more advantages than disadvantages as a platform for conducting reconnaissance, surveillance, and target acquisition.

Capabilities

To better understand the other advantages and disadvantages of this option, CBO compared the reconnaissance capabilities of an air cavalry troop made of up of 12 tilt-rotor UAV systems and one made up of 12 Comanche helicopters. In the comparison, CBO used a typical mission profile for Comanche: flying to a radius of 200 kilometers and operating there for 40 minutes. The comparison assumed that all 12 helicopters and UAV systems would be available at any given moment. By several different measures, the UAVs had a number of advantages over the Comanches (see Table 10).

With respect to loiter time—the amount of time each aircraft can stay aloft over an area—the tilt-rotor UAV has roughly a 6-to-1 advantage over Comanche. It can provide more than four hours of loiter time, compared with approximately 40 minutes for Comanche.[5] With respect to dash speed, the UAV is actually a little

---

4. Bell Helicopter has plans in which Eagle-Eye could carry and deliver sensor-fuzed weapons. But in analyzing this option, CBO did not consider those plans feasible in the short or intermediate term because the UAV is still at the demonstrator stage and has not yet flown many hours and because the military's concepts of operation for using combat UAVs are still relatively undeveloped.

5. That figure assumes that Comanche is not using its external fuel tanks, which it is not likely to do most of the time.

faster than the helicopter (370 kilometers per hour versus 315). Thus, it has a slightly greater ability to be in the right place at the right time.

Assuming that both the UAV and the Comanche troops flew out to a radius of 200 kilometers, both could observe 12 separate points or targets (one per platform) —something that resembles area reconnaissance. In making its comparison, CBO chose the most conservative assumption: substituting one UAV system with three air vehicles and a ground control station for one Comanche, rather than one air vehicle for one helicopter or some other trade-off. Observing multiple targets or areas simultaneously was deemed to be a crucial capability for a troop to have. In a real operation, of course, there may not always be 12 separate and distant targets to observe, or the 12 Comanche helicopters may operate separately.

TABLE 10.   PERFORMANCE CAPABILITIES OF A COMANCHE VERSUS
             A TILT-ROTOR UAV

	Comanche Reconnaissance Helicopter	Eagle-Eye Tilt-Rotor UAV
Radius (Kilometers)	200	200
Loiter Time (Hours)[a]	0.7	4
Dash Speed (Kilometers per hour)	315	370
Number of Targets or Spots Covered	1	1
Maximum Range (Kilometers)	500[b]	200[c]

SOURCE: Congressional Budget Office.

a.   The length of time the aircraft can stay aloft at its radius.

b.   Using internal fuel tanks. External tanks, which are not typically used, would extend the range by almost 2,000 kilometers.

c.   Limited by line-of-sight communications and data link.

Deployability

CBO did not conduct a detailed analysis of the deployability requirements of a cavalry squadron with two troops of Comanches versus one with a troop of Comanches and a troop of UAVs. Because cavalry squadrons contain a large, broad array of equipment other than helicopters (including tanks and artillery), the overall difference in the number of transport aircraft that the two types of units would need is likely to be marginal. That is particularly true since one Comanche and the air vehicles and mobile ground station of one tilt-rotor UAV system can each fit on one C-130. (The support equipment and the personnel to man and maintain both the helicopter and the UAV system would deploy separately.)

Overall Assessment

The primary effect of Option III would be to provide the Army with a substantial amount of reconnaissance capability at a much lower cost. The Army's reconnaissance helicopters, including Comanche, are chiefly intended to collect information. By spending extra money to acquire Comanche rather than a tilt-rotor UAV for that mission, the Army is getting a more versatile aircraft that can attack the enemy as well as observe it. However, it is also putting pilots at risk on each reconnaissance mission, even those that do not require a combat engagement.

Other factors could figure in any trade-off between unmanned aerial vehicles and reconnaissance helicopters. Helicopters are self-contained platforms in the sense that their pilots can see, report, and do whatever is necessary (in accordance with their commander's instructions). With UAVs, imagery is sent via a data link to a ground terminal where the commander can see it directly. But if the target under observation needs to be attacked, a second platform has to do the job, which can take time. In a critical situation, there may be no substitute for having an aircraft that can, for example, identify an advanced reconnaissance unit of the enemy and immediately destroy it.

UAVs, however, have the advantage of being able to operate in nuclear, biological, or chemical environments without jeopardizing the lives of their controllers. In that sense, UAVs are more versatile than helicopters. In addition, UAVs have the potential to carry many payloads other than video or infrared sensors. The payload capacity of the tilt-rotor UAV is large enough to accommodate signals intelligence packages, mine-detection equipment, and even potentially some types of armament.

OPTION IV: USE GLOBAL HAWK UAVs TO SUBSTITUTE FOR THE REDUCTION OF JSTARS

This option proposes taking greater advantage of unmanned aerial vehicles by purchasing a fleet of Global Hawks to give theater commanders more wide-area surveillance capability. That capability is now provided by the Joint Surveillance Target Attack Radar System (JSTARS)—a joint Army/Air Force reconnaissance system that combines a powerful multimode ground-surveillance radar with command-and-control systems on board a 707 aircraft. The purpose of the system is to detect mobile and stationary targets on the ground and transmit their locations to ground commanders and combat aircraft.

The Air Force had planned to buy 19 JSTARS aircraft in order to provide coverage for two combat theaters simultaneously. However, the recent Quadrennial Defense Review proposed reducing that purchase to 13 (plus one for testing). The Department of Defense argued that a fleet of 13 JSTARS would be able provide the round-the-clock coverage needed in a major theater war. In the event of a second conflict, some of the aircraft would have to redeployed to the second theater, possibly opening gaps in coverage. DoD plans to "explore the potential for supplementing radar coverage of enemy force movements from long-endurance unmanned aerial vehicles."[6] This option reflects that idea.

There are several ways to achieve the same capability as JSTARS using high-altitude endurance UAVs, specifically Global Hawk. For example, one study commissioned by DoD proposed building advanced Global Hawks that would take advantage of existing technology to incorporate a radar system as capable as the one aboard JSTARS. CBO did not explore that option in great detail because such a choice assumes even greater technical risk than already exists in the development of Global Hawk. The current Global Hawk has been designed and built with one engine and has had only very limited testing so far. An advanced Global Hawk along the lines of the one in DoD's study would require two engines, substantial new development, and all of the accompanying technical risk that those entail.

A second alternative would be to use the Global Hawk as currently configured but to achieve the same capability as JSTARS by substituting processing power for electrical power and providing a more advanced radar.

The third alternative—which is the one that CBO explored in this option—is to provide the capability of a JSTARS by using several of the currently configured Global Hawks and their radars in the moving-target-indicator mode. According to

---

6.   Secretary of Defense William S. Cohen, *Report of the Quadrennial Defense Review* (May 1997), p. 45.

CBO's analysis, that would require having at least three Global Hawks aloft simultaneously.

Costs

Assuming that the Air Force follows the Quadrennial Defense Review's recommendation to reduce the JSTARS fleet from 19 to 13 aircraft, this option would cost almost $1.2 billion more than the Air Force's plan in acquisition and 15-year operating and support costs. The Global Hawks bought under this option would be in addition to the planned purchases of Global Hawks for other reconnaissance and surveillance missions as well as the 13 JSTARS aircraft.

The reduction proposed by the Quadrennial Defense Review assumed that the NATO alliance would purchase a number of JSTARS, which would supplement the U.S. fleet in times of crisis, if necessary. But NATO has decided not to acquire those aircraft. As a result, many people inside and outside the Congress are suggesting that the Air Force buy the original 19 aircraft after all. In that event, this option would substitute 11 Global Hawks for the extra six JSTARS aircraft, saving a total of around $2.3 billion in acquisition and 15-year operating and support costs.

Capabilities

The radar systems in both JSTARS and Global Hawk include a moving-target indicator that detects moving vehicles. There are important differences, however, in the capability of those indicators. The major measures of their capability are ground-referenced coverage area, revisit (or update) time, minimal detectable velocity, range resolution, and azimuth accuracy (see Table 11).

Ground-Referenced Coverage Area. The ground-referenced coverage area is the area on the ground that the radar's moving-target indicator can cover. According to unclassified sources, that area for JSTARS is 150 kilometers by 180 kilometers, or 27,000 square kilometers—the notional area for which an Army corps has responsibility. (JSTARS's actual coverage capability is larger, although the exact figure is classified.)[7] The ground-referenced coverage area of Global Hawk's radar, as it is currently configured, is about one-third of that. Thus, providing the same area coverage as one JSTARS would appear to require at least three Global Hawks.

---

7.   John Haystead, "JSTARS—Real-Time Warning and Control for Surface Warfare," *Defense Electronics* (July 1990), p. 39.

TABLE 11.   COMPARISON OF THE CAPABILITIES OF THE MOVING-TARGET
                INDICATORS ON JSTARS AND GLOBAL HAWK

	Global Hawk	JSTARS
Ground-Referenced Coverage Area (Square kilometers)	About 9,500	More than 27,000
Revisit Time (Seconds)	70	At least 60
Minimum Detectable Velocity (Knots)	4	a
Range Resolution (Meters)	10 to 20	a
Azimuth Accuracy (Meters)	350	Less than 350

SOURCE: Congressional Budget Office based on data from the Defense Advanced Research Projects Agency.

NOTE: JSTARS = Joint Surveillance Target Attack Radar System.

a.   Classified.

Substituting three Global Hawks for a JSTARS could create some command-and-control problems, however. For example, it would require more coordination and possibly better data links with the units on the ground who would be receiving the images from the UAVs. Because three aircraft would be doing the job of one and there would necessarily be some overlap in coverage area, ground units might have to access the imagery from all three Global Hawks that were aloft and know which air vehicle was covering which part of the ground.

Revisit Time. The revisit (or update) time is the amount of time required for the radar to sweep over an area and provide the latest available imagery of moving targets. The revisit time for Global Hawk's radar over the ground-referenced coverage area is around 70 seconds. JSTARS has a faster revisit time—at least 60 seconds—although the actual number is classified.[8]

Revisit time is important for several reasons. The moving-target indicator only detects vehicles that are moving at a certain minimum speed. The radar image appears as lighted dots on a screen. (Tracked vehicles and wheeled vehicles can be displayed as two different colors.) The average person will not be able to make much sense of a moving-target-indicator image other than to know that there are moving vehicles within the sweep of the radar. A trained expert in such imagery, however,

---

8.   Joris Janssen Lok, "Joint STARS Gains vs Greater Radar Control," *Jane's Defence Weekly,* April 23, 1997, p. 30.

will be able to distinguish among the dots and often determine the size and type of the units. If the delay between sweeps of the radar is too great, it is more difficult to determine which dot is which from the previous sweep. Thus, tracking particular formations of vehicles may be more difficult if the lag time is too great.

In addition, because both JSTARS and Global Hawk are equipped with synthetic aperture radars, which can provide an image of a particular location, it is important to know when a vehicle or column of vehicles has stopped moving. When that happens the vehicles disappear from the moving-target indicator. If they are sufficiently interesting targets, however, a radar operator can use the synthetic aperture radar to determine what they are. But too long a lag time between sweeps of the radar may make it more difficult for the operator to find the target with the synthetic aperture radar after it has disappeared from the moving-target indicator.

The key issue is whether Global Hawk's slower revisit time is quick enough for the moving-target indicator to provide useful imagery. That appears to be the case, particularly because today the use of such indicators is geared toward detecting large formations of vehicles. A revisit time of 70 seconds will probably not have much effect on an analyst's ability to determine what vehicle formations the radar is picking up, although clearly it will have a slight negative impact (relative to JSTARS's revisit time). In the future, if moving-target indicators are used to try to identify more specific targets or vehicles, Global Hawk's radar may be at a disadvantage. But by then, there may also be more advanced radars with faster revisit times that Global Hawk could use.

Minimum Detectable Velocity. As its name suggests, minimum detectable velocity is the speed at which an object must be traveling to be detected by the moving-target indicator. However, only objects that are moving with some degree of perpendicularity to the radar signal will be picked up. Objects that are moving strictly parallel will not be detected. The reason is that as an object moves toward or away from the radar, it causes a shift in the radar signal and thus is detected. If an object is moving parallel to the radar, there is no shift in the signal or the shift is so slight that the object appears stationary—or, in effect, as though it were not there.

The minimum detectable velocity for JSTARS is classified. But for Global Hawk, that speed is four knots if the target is moving directly toward or away from the radar. If the object is traveling at an angle to the radar, it must be going at a higher speed to have the effect of moving toward or away from the radar at four knots. For example, at a slight angle from the perpendicular, an object may need to be moving at six to 10 knots to register. At a sharper angle, the speed would have to be much greater for the object to be detected.

Range Resolution. Another aspect of a radar's moving-target-indicator mode is the degree to which it can distinguish between two objects. In the most literal sense, a radar must "resolve" the imagery it receives. Different radars can do that to different extents, just as two people looking at an optometrist's eye chart may have different abilities to distinguish between the letters on the chart. If a convoy of vehicles is traveling along a road, those vehicles must be a certain distance apart for the radar to recognize them as separate objects. The closer together they can be and still be detected separately, the better range resolution the radar has.

Global Hawk's moving-target indicator is particularly good at resolving objects that are short distances apart, about 10 to 20 meters. An opponent would probably have a difficult time bunching its vehicles so close together that the radar would be unable to distinguish between them. The value of a good range resolution is most pronounced when a trained analyst of moving-target-indicator imagery is attempting to determine whether a particular set of vehicles represents a platoon, a company, or some other size unit. Being able to count—even roughly—the number of vehicles in an enemy unit may provide valuable information to a battlefield commander.

Azimuth Accuracy. Azimuth accuracy is relatively straightforward: it is how precisely a radar can determine the location of the objects it detects. By this measure, Global Hawk's moving-target indicator clearly has less capability than that of JSTARS. Global Hawk's indicator can place the location of a vehicle up to 350 meters away from its actual position, whereas JSTARS is expected to be more accurate. However, that weakness is mitigated somewhat by the fact that moving-target indicators are mostly concerned with identifying relatively large formations of vehicles rather than tracking individual vehicles. Thus, if the indicator picks up a convoy of vehicles traveling in a line 200 meters parallel to a road, the analyst can assume that the formation is probably on the road and that the 200-meter difference is within the radar's margin of error.

Other Capability Issues. A significant advantage of Global Hawk over JSTARS is that its moving-target indicator can provide "deep" (that is, long-distance) coverage without risking the lives of an aircrew. JSTARS is intended to operate at the forward line of U.S. troops and provide coverage over a range of 160 kilometers. At that distance, however, it would not detect whether enemy reinforcements were entering the theater and what kind of forces they were. To give a specific example, a JSTARS operating near the Kuwaiti border with Iraq would not be able to indicate the size and scope of any forces that might be moving into the theater from Basra or beyond. But Global Hawk, which is unmanned, could do so without risking an aircrew. Attacking enemy forces with precision munitions deeply (long before they are able to reach the forward line of U.S. troops) is an important element of future U.S. warfighting strategy. For that reason, it might make sense to buy even more Global Hawks to provide deep coverage or use some of the ones intended for coverage at the forward line of U.S. troops for deep coverage.

Deployability

In general, JSTARS and Global Hawk aircraft would self-deploy (fly to the theater of operations) in the event of a major regional conflict. But the exact deployment requirements of the Air Force's plan and Option IV are unclear, in part because it is unclear what the final JSTARS and Global Hawk programs will look like.

Overall Assessment

Assuming the Air Force follows the recommendation of the Quadrennial Defense Review to reduce the number of JSTARS aircraft, Option IV would give one or more theater commanders substantial additional capability to detect moving targets at a total cost of $1.2 billion. If instead the Air Force buys additional JSTARS aircraft, Option IV would save money in comparison. Global Hawk is not quite as capable as JSTARS, but it may be good enough for its intended missions and can be used without risking aircrews. Probably the most serious drawback of Option IV is the additional command-and-control efforts that would be required to make the option work.

OPTION V: END DARKSTAR AFTER THE ACTD AND RELY ON OTHER SYSTEMS

As noted in Chapter I, some Members of Congress have raised concerns about apparent overlaps between three of DoD's unmanned aerial vehicle programs: Darkstar, Global Hawk, and Predator. This option would address those concerns by cancelling the Darkstar program after its Advanced Concept Technology Demonstration phase ends and relying on other types of endurance UAVs instead. Darkstar is a high-altitude UAV that is expected to have low-observable (stealthy) characteristics. It is designed to carry out a particular mission: collecting imagery over highly defended targets before an enemy's air defenses have been suppressed. In addition, it may be particularly useful in supporting special-operations forces.

Aside from its stealthiness, Darkstar is expected to be less capable than Global Hawk, although more capable, for the most part, than Predator (see Table 12). The Defense Airborne Reconnaissance Office and the Air Force have described Global Hawk as a highly capable but moderately survivable UAV, whereas Darkstar

TABLE 12.   COMPARISON OF THE CAPABILITIES OF PREDATOR,
             DARKSTAR, AND GLOBAL HAWK

	Predator	Darkstar	Global Hawk
Maximum Range (Kilometers)[a]	4,200	5,600	25,500
Operating Altitude (Feet)	10,000-25,000	40,000-45,000	55,000-65,000
Cruise Speed (Kilometers per hour)	120	463	639
Endurance at Radius	20 hours at 926 km	8 hours at 926 km	22 hours at 5,556 km
Payload (Pounds)	450	1,000	2,000
Survivability	?	?	?

SOURCE: Congressional Budget Office based on data from the Department of Defense.

NOTE: km = kilometers.

a.   Maximum range is the farthest the unmanned aerial vehicle (UAV) can fly before running out of fuel. It differs from
      radius in that the tactical UAVs are limited to an operating radius far short of the range of the air vehicle because they
      communicate through line-of-sight links. Endurance UAVs are not limited to a particular radius because, when not
      in an autonomous mode, they communicate with their controllers by satellite.

is a highly survivable but moderately capable UAV. The chief advantage of buying Darkstar, therefore, is to buy stealthy reconnaissance capability.

The degree to which Darkstar really is a highly stealthy imagery-collection platform, however, is not clear—especially compared with Global Hawk and Predator. According to DoD, tests in May 1996 "validated" Darkstar's low-observable design.[9] But the Darkstar program is still recovering from the crash of the first air vehicle, and much work remains before it will be successfully completed. A true test of Darkstar's stealth, including when its sensors are engaged, must wait until further development and military-utility demonstrations take place.

Although not stealthy, Global Hawk and Predator have some features that might help them survive in an environment in which enemy air defenses had not been suppressed. (For more on the issue of UAVs' survivability, see Box 2.) Global

---

9.   Department of Defense, Defense Airborne Reconnaissance Office, *UAV Annual Report, FY 1996*
      (November 6, 1996), pp. 22-23.

BOX 2.
UAV SURVIVABILITY IN WARTIME

How capable will unmanned aerial vehicles (UAVs) be of surviving a war, and does it matter? Military officials are likely to disagree. On the one hand, some may believe it would not matter if UAVs suffered a high attrition rate in wartime because they are less expensive than other weapons and reconnaissance platforms, and lives are not lost when they are shot down. UAVs were built to take on risky reconnaissance assignments and meant to suffer attrition. Tactical UAVs in particular are intended to be quite inexpensive—generally less than $1 million per air vehicle, with the cheapest ones expected to cost less than half that much. Thus, if they were subject to enemy fire, particularly missile fire, that means there would be fewer shots being aimed at manned aircraft.

On the other hand, some officials might argue that although UAVs are relatively inexpensive, they are not cheap, so they should still avoid attrition. Furthermore, some analysts expect UAVs to become so important to the various U.S. commanders involved in a military operation that those commanders will be unwilling to risk them in situations where attrition is likely. That could be more likely with the high-endurance UAVs that would be assigned to a theater commander than with the tactical UAVs—although even that might depend on the quantity and availability of such UAVs.

Until unmanned aerial vehicles are deployed with U.S. forces and used in a large-scale military operation or war, no one knows for sure how survivable they will be. Tactical UAVs are expected to be less survivable than endurance UAVs, primarily because they fly around the battlefield and are in the thick of the fighting. Still, their survivability could well be higher than the common perception. Tactical UAVs fly relatively slowly, which could enhance their survivability against tactical surface-to-air missiles (SAMs) because it would allow them to get lost in the clutter of the battlefield. Furthermore, because those UAVs are smaller than any manned platform, their radar signature will probably be smaller as well.

According to a Marine Corps briefing, the Pioneer UAVs that were used in the Gulf War were small enough and, most of the time, flew at high enough altitude that the enemy could not see them. As a result, although Pioneers flew more than 150 sorties during the war, only a handful were shot down.[1] During the Army's Advanced Warfighting Experiment at the National Training Center in early 1997, a Hunter UAV in support of brigade and division operations was "shot down" by a simulated SAM engagement, but only after it had loitered over the target for two hours.[2] However, several Predator UAVs have been lost in Bosnia, some to enemy fire.

What might attrition for tactical UAVs look like if the rate was considerably higher than in peacetime or even in the Gulf War? (Wartime attrition will almost certainly be higher than peacetime attrition, and Iraq was a particularly inept opponent.) Assume, for example, that 21 brigades—the equivalent of seven divisions—deployed to a major regional conflict and that each brigade had the equivalent of one Outrider tactical UAV system. If the wartime attrition rate was four times worse than the expected peacetime attrition rate, almost the entire UAV force with those 21 brigades would be destroyed in 45 days.

The endurance UAVs—Global Hawk and Darkstar—should be the most survivable unmanned aerial vehicles once they are deployed and used. Darkstar is a low-observable air vehicle designed to penetrate enemy air defenses, record imagery, and return. Global Hawk is designed to fly around 60,000 feet and carry a suite of countermeasures to confuse enemy SAMs. Nevertheless, Global Hawk is not intended to penetrate enemy air defenses. In the case of an actual war, the Department of Defense expects to use it primarily after enemy air defenses have been destroyed.

---

1.   Department of Defense, *Conduct of the Persian Gulf War* (April 1992), p. 723. Total losses were 12 destroyed and 14 to 16 damaged—mostly because of electromagnetic interference with their data and communication links and human error as a result of fatigue.

2.   "Hunter UAV Gets High Marks at Army AWE Despite Flight Anomaly," *Aerospace Daily*, March 28, 1997, p. 466.

Hawk flies higher than all but the most capable of surface-to-air missiles (SAMs). Furthermore, it is equipped with an electronic countermeasures suite to thwart enemy SAMs in the event that they target the air vehicle. Predator has an all-composite air frame (which produces a smaller radar reflection than an air frame made of metal), can operate at night, and flies slower than enemy air-defense radars are typically programmed to detect.

Does that mean Global Hawk and Predator are as survivable as Darkstar? Probably not. It does suggest, however, that in light of the less threatening environment that the United States faces for the foreseeable future and the "niche" mission that Darkstar fulfills, it may be cost-effective to end the Darkstar program at the conclusion of its development process. Three air vehicles are expected to be left over from the Advanced Concept Technology Demonstration. They would still be available for the rare, high-value mission in which having a stealthy reconnaissance collector was especially desirable. Global Hawk and Predator could also be used for those infrequent missions if commanders were willing to accept the risk of higher attrition or the possibility that the imagery might not be collected.

Moreover, the types of systems that would probably pose the greatest threat to Global Hawk—the SA-10s and SA-12s, or "double-digit SAMs"—are relatively rare. And they would most likely be one of the first targets of U.S. air forces seeking to establish air superiority. Once those SAMs were destroyed, Global Hawk could operate more freely even if the rest of the enemy's air-defense systems had not yet been suppressed.

Not all Air Force officials or defense analysts would agree with the analysis underpinning this option. Some argue that the United States will face severer threats in the future because of the expected proliferation of high-quality surface-to-air missiles. Furthermore, one Air Force official argues, with the retirement of the SR-71 high-altitude reconnaissance plane, the United States no longer has the means to collect imagery over highly defended targets. Satellites cannot always be relied on to be in the right place at the right time, and other reconnaissance aircraft, such as the U-2, can be shot down by SAMs.[10] Global Hawk and Predator will also be vulnerable.

---

10. It is worth noting that according to *Defense News*, the Administration has stated that there is no military requirement for the SR-71, a position the Air Force supports. See William Matthews, "USAF Again Retires SR-71 Reconnaissance Planes," *Defense News*, May 18, 1998, p. 32.

Costs

Because this option would not buy anything to replace Darkstar, the savings stem directly from ending the program. Savings in acquisition costs alone would be more than $600 million compared with the Air Force's likely plan, plus another $400 million in 15-year operating and support costs.

Overall Assessment

In light of the fact that Darkstar does not provide any additional capability over Global Hawk besides stealth, the price of Darkstar is in fact the price of stealth in the UAV force structure. The Congress must decide how much it wants to pay for stealthy UAVs; right now, the price tag appears to be about $1 billion. Furthermore, the mission that Darkstar would fulfill is likely to be infrequent and could be accomplished by the three ACTD prototypes or by Global Hawk or Predator when necessary.

www.ingramcontent.com/pod-product-compliance
Lightning Source LLC
Chambersburg PA
CBHW081217170526
45165CB00009B/2856